THEORY AND APPLICATION OF DRILLING FLUID HYDRAULICS

The EXLOG Series of Petroleum Geology and Engineering Handbooks

Theory and Application of Drilling Fluid Hydraulics

Written and Compiled by
EXLOG Staff

Edited by Alun Whittaker

International Human Resources Development Corporation • Boston

© 1985 by EXLOG®.* All rights reserved. No part of this book may be used or reproduced in any manner whatsoever without written permission of the publisher except in the case of brief quotations embodied in critical articles and reviews. For information address: IHRDC, Publishers, 137 Newbury Street, Boston, MA 02116.

Library of Congress Cataloging in Publication Data

Main entry under title:

Theory and application of drilling fluid hydraulics.

Includes bibliographies and index.
1. Drilling muds. I. Whittaker, Alun. II. EXLOG (Firm).
TN871.2.T428 1985   622'.338   84-25172
ISBN 0-88746-045-3

Printed in the United States of America

*EXLOG is a registered service mark of Exploration Logging Inc., a Baker Drilling Equipment Company.

CONTENTS

List of Illustrations ix
Preface xi

1. INTRODUCTION 1
FUNDAMENTAL PROPERTIES OF FLUIDS 1
DENSITY 2
HYDROSTATIC PRESSURE 2
PASCAL'S LAW 4
REFERENCES 5

2. FLUID FLOW: PRINCIPLES, MODELS, & MEASUREMENT 7
FLUID-FLOW PRINCIPLES 7
   FLUID DEFORMATION 8
   FLOW REGIMES 10
      Plug Flow 10
      Laminar Flow 10
      Turbulent Flow 13
      Regime Determination 13
   CONTINUITY OF FLOW 14
FLUID MODELS 16
   INTRODUCTION 16
      Newtonian Fluid Model 16
      Bingham Plastic Model 17
      Power Law Fluid Model 19
      Other Models 21
   TIME-DEPENDENT BEHAVIOR 22
MEASUREMENT OF FLUID FLOW PROPERTIES 24
   INTRODUCTION 24
   INSTRUMENTS 24
   ANALYSIS 26
   RESULTS 36
   FIELD PROCEDURES 38
REFERENCES 44

3. THE DRILLING FLUID 47
INTRODUCTION 47
WATER-BASED MUDS 47
OIL-BASED MUDS 49
MUD VISCOSITY 49
REFERENCES 50

4. THE MUD CIRCULATING SYSTEM 51
INTRODUCTION 51
THE PUMPS 51
THE DRILLSTRING 55

THE BIT 55
THE ANNULUS 56
REFERENCES 58

5. THE DRILLSTRING 59
INTRODUCTION 59
LAMINAR FLOW 59
    ANALYSIS 59
    RESULTS 65
    FIELD PROCEDURES 66
    THE REYNOLDS NUMBER 68
CRITICAL VELOCITY 70
TURBULENT FLOW 70
TRANSITIONAL FLOW 76
TOOL JOINTS 78
REFERENCES 80

6. MOTOR, TURBINE, & BIT 83
INTRODUCTION 83
THE MOTOR 83
THE TURBINE 85
THE BIT NOZZLES 86
REFERENCES 88

7. THE ANNULUS 91
INTRODUCTION 91
LAMINAR FLOW 91
    EXACT SOLUTIONS 91
    APPROXIMATE SOLUTIONS 97
    PRACTICAL METHODS 100
        Newtonian Fluid 100
        Bingham Fluid 100
        Power Law Fluid 102
        Casson Fluid 103
        Robertson-Stiff Fluid 103
TURBULENT FLOW 103
TRANSITIONAL FLOW 105
REFERENCES 106

8. SWAB AND SURGE 111
GENERAL 111
REFERENCES 107

9. CUTTINGS TRANSPORT 119
GENERAL 119
REFERENCES 117

10. OPTIMIZING THE HYDRAULICS PROGRAM   129
GENERAL   129
REFERENCES   134

Appendix A: Nomenclature   137
Appendix B: Units   143
Appendix C: Key Equations   145
Appendix D: Example Hydraulics Calculations   167
Glossary   195
Index   199

ILLUSTRATIONS

2-1   Deformation of a Fluid by Simple Shear   8
2-2   The Dependence of the Shear Rate, q, upon the Distance of Planar Separation, dy   9
2-3   Three-Dimensional View of Laminar Flow in a Pipe for a Newtonian Fluid   11
2-4   Two-Dimensional Velocity Profile of Laminar Flow in a Pipe, for a Newtonian Fluid   11
2-5   Two-Dimensional Velocity Profile of Turbulent Flow in a Pipe, for a Newtonian Fluid   13
2-6   Continuity of Flow: Fluid Velocity is Inversely Proportional to the Cross-Sectional Area of the Fluid Conductor   15
2-7   Flow Curve for a Newtonian Fluid   17
2-8   Flow Curve for Bingham Plastic Fluid   18
2-9   Flow Curves for Power Law Fluid   20
2-10 Typical Drilling Fluid Vs. Newtonian, Bingham and Power Law Fluids   20
2-11 Gel Strengths   23
2-12 Marsh Funnel   25
2-13 Rotating-Sleeve Viscometer   26
2-14 Torque Acting on the Bob   28
2-15 Forces in Equilibrium on Rotating-Sleeve Viscometer   29
2-16 Displacement Between Two Fluid Shells Moving at Different Angular Velocities   31
2-17 K-Correction Factor   41
2-18 Typical Drilling Fluid Vs. Newtonian, Bingham and Power Law Fluids   43
2-19 Casson Fluid Intercept on the Shear Stress Axis   43
3-1   Ionic Interaction of Sodium Montmorillonite and Water   48
4-1   Rig Circulating System   52
5-1   Forces in Equilibrium for Laminar Pipe Flow   60
5-2   Approximation Error for Bingham Fluid   67
5-3   Relative Roughness of a Pipe   71
5-4   Friction Factor Curves for Newtonian Fluid, from the Colebrook Equation   73
5-5   Blasius Correlation for Various Power Law Fluids; $e/P = 0$ 75
5-6   Von Karmann Correlation for Various Power Law Fluids; $e/D = 0$ 75
5-7   Critical Reynolds Number for Pipe Flow, by Method of Hanks and Pratt   78
6-1   Moineau Motor   84
6-2   Turbine Motor   85
7-1   Velocity Profile of Laminar Flow in an Annulus, for a Newtonian Fluid   92
7-2   Velocity Profile of Laminar Flow in an Annulus, for a Fluid with a Yield Stress   92
7-3   Forces in Equilibrium for Annular Pipe Flow   94
7-4   Narrow-Annulus (Parallel Plate) Model   97
7-5   Narrow-Annulus Approximation, Power Law Fluid (Percentage Error in Calculated Pressure Drop)   99

7–6	Narrow-Annulus Approximation, Bingham Fluid (Percentage Error in Calculated Pressure Drop)	99
7–7	Approximation Error for Annular Flow of Bingham Fluid	102
7–8	Critical Reynolds Number for Annular Flow by Method of Hanks and Pratt	106
8–1	Equal Level Displacement	112
8–2	U-Tube Analogy for Equal Pressure Displacement	113
8–3	Clinging Constants for Bingham Fluid	115
8–4	Clinging Constants for Power Law Fluid	116
8–5	Clinging Constant for Turbulent Flow	117
9–1	Cuttings Transport	120
9–2	Walker and Mayes' Correlation for Drag Coefficient	121
9–3	Moore's Correlation for Drag Coefficient	122
9–4	Proposed Correlation for Drag Coefficient	123
10–1	Hydraulic Power Vs. Flowrate	131
10–2	Optimum Impact Force	133
D–1	Example—Condensed Mud Hydraulics	168
D–2	Example—Condensed Mud Hydraulics (Continued)	169
D–3	Example—Bit Hydraulics Optimization	170

PREFACE

The objectives of this book are (1) to serve as a reasonably comprehensive text on the subject of drilling hydraulics and (2) to provide the field geologist with a quick reference to drilling hydraulics calculations. Chapter 1 introduces the basic principles of fluid properties, and Chapter 2 presents the general principles of fluid hydraulics. Chapters 3 through 10 analyze specific hydraulic considerations of the drilling process, such as viscometric measurements, pressure losses, swab and surge pressures, cuttings transport and hydraulic optimization. References are presented at the end of each section.

The units and nomenclature are consistent throughout the manual. Equations are given generally in consistent S.I. units; some common expressions are also given in oilfield units. Nomenclature is explained after every equation when necessary, and a comprehensive list of the nomenclature used is given in Appendix A. Units are listed in Appendix B. In Appendix C, all the important equations are given in both S.I. and oilfield units. Appendix D contains example hydraulics calculations. A glossary is included.

THEORY AND APPLICATION OF DRILLING FLUID HYDRAULICS

1
INTRODUCTION

To drill a well safely and successfully depends upon a thorough understanding of drilling hydraulics principles. Thus, drilling hydraulics is a very important subject with which all logging geologists should be familiar.

Drilling hydraulics is, in part, a specific application of the principles of fluid mechanics -- the study of the behavior of fluids. The principles of fluid mechanics are derived from the principles of Newtonian mechanics for solid bodies; differences are largely a matter of interpretation due to the nature of the fluid medium. Hence, a discussion of the fundamental nature of a fluid provides a starting point for drilling hydraulics analysis.

1.2   FUNDAMENTAL PROPERTIES OF FLUIDS

We classify matter into two categories -- solids and fluids. Solids are rigid; fluids flow. However, the distinction is not always clear-cut. Some substances may act as both a solid and a fluid. A thorough understanding of drilling hydraulics must start with an investigation into the fundamental nature of fluids -- that is, the properties which distinguish a fluid from a solid. These properties, which are responsible for the main phenomena of drilling hydraulics, are analyzed and quantified throughout this manual.

There are two classes of fluids: gases and liquids. Gases are highly compressible; that is, volume is dependent upon pressure. Additionally, the volume of a gas is dependent upon temperature. Liquids, on the other hand, are only slightly compressible; that is, volume varies only a small amount with changes in pressure. Additionally, the volume of a liquid does not depend significantly upon temperature.

This manual analyzes drilling hydraulics for liquids only. However, because drilling muds are generally called "drilling fluids," the term "fluid" is used throughout the text. For the applications discussed in this manual, the volume of the drilling fluid is not significantly altered by changes in pressure and temperature; therefore, the effects of pressure and temperature on volume are ignored in the analyses.

The inability of a fluid to maintain a rigid shape is a very important characteristic. Consider a block of wood set upon a table. It is possible to push on the side of the block, sending it sliding across the tabletop. Now, consider a hypothetical "block" of water set upon a table. It would be impossible to move the block as a whole by pushing on its side. A fluid cannot sustain a shear stress. A force may be applied to a solid from any direction, but for a fluid at rest, a force can only be applied perpendicularly to the surface. A tangential force will cause a fluid to lose its shape, to constantly deform, to flow.

1.2 DENSITY

A cube of water measuring 1 foot along each edge weighs 62.4 pounds. Thus the weight density, or specific weight, is 62.4 lb/ft^3. Specific weight divided by the gravitational constant is called "mass density," or just "density." In the oilfield, the measured mudweight of a drilling fluid is the weight density, usually given in units of lb/gal.

1.3 HYDROSTATIC PRESSURE

Hydrostatic pressure is the pressure which exists due to the weight of a static column of fluid. The weight of a cube of water, measuring 1 foot along each edge, is distributed evenly over the bottom surface of an area 1 ft^2, and exerts a hydrostatic pressure of 62.4 lb/ft^2 (0.433 psi).

The hydrostatic pressure of a column of drilling fluid is given by:

$$P_h = (D_v - Fl) * \rho * g \qquad (1-1)$$

where

P_h = hydrostatic pressure
D_v = vertical depth
Fl = flowline depth
ρ = fluid density
g = gravitational constant

or, in oilfield units

$$P_h = (D_v - Fl) * MW * .0519$$

where

P_h = hydrostatic pressure (psi)
D_v = vertical depth (ft)
Fl = flowline depth (ft)
MW = mudweight (lb/gal)

The number 0.0519 is a conversion factor for oilfield imperial units (psi, lb/gal/ft) and is derived as follows:

There are 7.48 gallons in 1 cu ft
There are 144 sq inches in 1 sq ft

hence

$$lb/gal * 7.48 \ gal/ft^3 * \frac{1}{144} \ ft^2/in.^2 = psi/ft$$

therefore

$$\frac{7.48}{144} = psi/ft/lb/gal$$

$$0.0519 = psi/ft/lb/gal$$

So fresh water, having a density of 8.34 lb/gal, or 62.4 lb/cu ft, exerts a pressure of

$$8.34 * 0.0519 = 0.433 \text{ psi/ft}$$

Similarly, using S.I. units:

$$P(kPa) = W(kg/m^3) * D(m) * 0.0098$$

1.4 PASCAL'S LAW

Consider a block of water to be submerged in a lake. Although the block exerts a hydrostatic pressure at its base, the block remains stationary. The reason is that at any point along the surface of the block there are two equal but opposite forces -- the force of the block pushing out and the force of the surrounding water pushing in.

This principle is generalized by Pascal's Law:

> "The pressure at any point in a static fluid is the same in all directions. Any pressure applied to a fluid is transmitted undiminished through the fluid."

The second part of Pascal's Law has a very important consequence in drilling hydraulics. When a well is shut in during a kick, the excess pressure of the kicking formation is transmitted undiminished through the fluid. The weaker formations near the surface experience the same increase in pressure as the better consolidated formations downhole. The weaker formations uphole are the most likely to fracture and cause loss of circulation.

1.5   **REFERENCE**

1. Henke, Russell W., <u>Introducton to Fluid Mechanics,</u> Addison-Wesley Publishing Co., Reading, MA, 1966.

2
FLUID FLOW: PRINCIPLES, MODELS, & MEASUREMENT

2.1   FLUID-FLOW PRINCIPLES

As emphasized in Section 1, the inability of a fluid to sustain a tangential force is a very important characteristic. A tangential force will cause a fluid to lose its shape or to deform. Continuous deformation is known as "flow."

Fluid flow must always be considered to take place within some conductor. The conductor need not be as obvious as a cylindrical pipe. A fluid flowing down an inclined tabletop is enclosed by the table surface underneath and the atmosphere on the sides and top.

The nature of the conductor greatly affects the flow behavior of a fluid. All analysis of fluid flow must take into account the geometrical shape of the conductor. Thus in this manual, fluid flow down the drillstring is analyzed separately from fluid flow up the annulus. The necessary consideration of conduit geometry is largely responsible for the mathematical complexity of analysis.

In general, fluid flow is the result of parallel fluid layers sliding past one another. The layer adjacent to the surface of the conductor adheres to the surface, and each successive layer slides past its neighbor with increasing velocity. This orderly means of flow is called "laminar flow."

At higher average velocities the laminae lose their order and crash randomly into one another, and orderly flow remains only very close to the conductor surface. This type of flow is called "turbulent flow."

2.2 FLUID DEFORMATION

The deformation of an element of fluid is illustrated in Figure 2-1. Two parallel fluid layers are separated by a distance dy. An applied force, F, acting over an area A, causes the layers to slide past one another. The resistance to this sliding movement, the frictional drag, is called "shear stress."

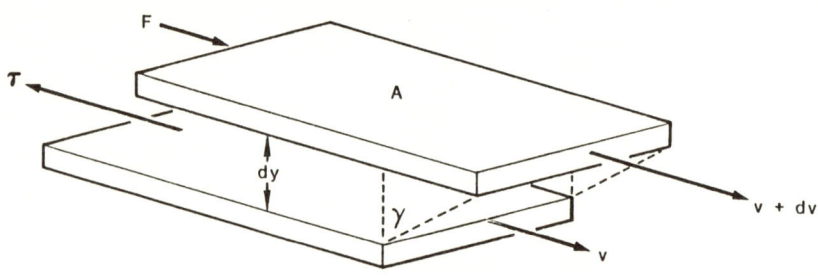

Figure 2-1. Deformation of a Fluid by Simple Shear

The shear stress, τ, is commonly defined as the applied force F divided by the area over which it acts, A, or $\tau = \frac{F}{A}$. Because the shear stress is commonly interpreted to be acting in a direction opposite to the direction of the applied force F, the given definition of shear stress may cause some confusion. The confusion is resolved by understanding that the deformation model represents a system in equilibrium. The velocities of the fluid layers, v and v + dv, are constant. A constant velocity indicates that no net force is acting upon the fluid layers to cause a change in the velocities or an acceleration. Because the net force equals zero, the applied force, F_1, is equal in magnitude and opposite in direction to the frictional forces. Thus shear stress may be defined in terms of the applied force, F.

When a force is initially applied to a static fluid, the fluid accelerates from a velocity of zero until it reaches some constant average velocity. During this time, the applied force F is greater than the frictional forces, and the net force accelerates the fluid. The state of equilibrium may be considered to be the state of the system after a very long or infinite time of flow. Fluid flow analysis considers the system under analysis to be in equilibrium.

The magnitude of shear between the two layers is represented by the value γ, the shear rate. The shear rate is defined as the difference in the velocities between the two layers divided by the distance of separation, or $\gamma = \frac{dv}{dy}$. The reason for dividing by the distance of separation dy may not be apparent. Figure 2-2 illustrates the change in γ that is due to only a change in dy. The distance d represents the distance that the top fluid layer has traveled after a set amount of time.

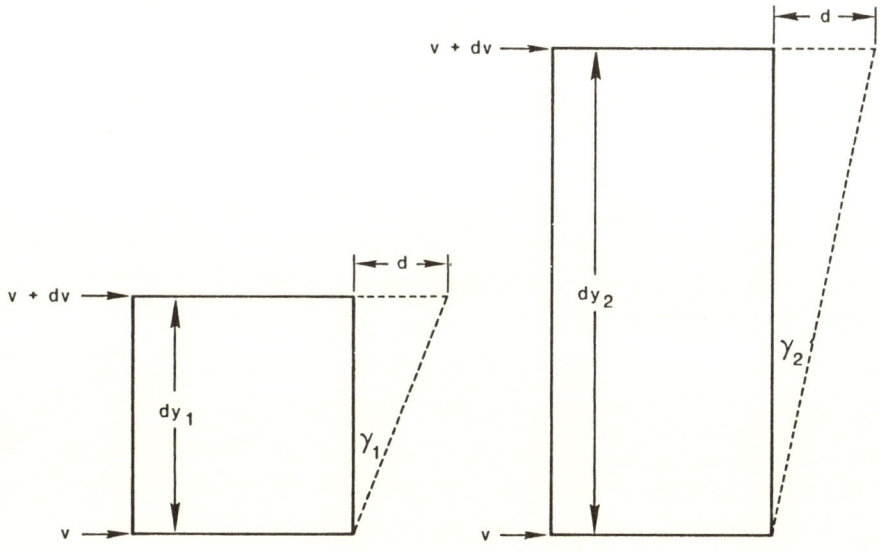

Figure 2-2. The Dependence of the Shear Rate, γ, upon the Distance of Planar Separation, dy

The relationship between shear stress, τ, and shear rate, γ, defines the flow behavior of a fluid. For some fluids, the relationship is linear that is, if the shear stress is doubled, then the shear rate will also double. Such fluids are called Newtonian fluids. Most drilling fluids are non-Newtonian fluids. They are defined by a more complex relationship between shear stress and shear rate.

2.3 FLOW REGIMES

2.4 Plug Flow

In plug flow the fluid moves essentially as a single, undisturbed solid body, or plug. Movement is made possible by the slippage of a thin layer of fluid along the conductor surface. Plug flow occurs generally at very low flowrates.

2.5 Laminar Flow

The laminar flow of a Newtonian liquid in a circular pipe is illustrated in Figure 2-3. The laminae are concentric cylindrical shells which slide past one another like sections of a telescope. The velocity of the shell at the pipe wall is zero, and velocity of the shell at the center of the pipe is maximum.

A two-dimensional velocity profile is illustrated in Figure 2-4. Notice that the shear rate, defined as $\frac{dv}{dr}$, is simply the slope of a line at any point along the velocity profile. The shear rate is maximum at the wall and 0 at the center of the pipe. Since for a Newtonian fluid the shear stress is directly proportional to the shear rate, the shear stress is also maximum at the pipe wall and zero at the center of the pipe. Thus, both shear rate and shear stress vary with radial distance within the pipe.

The concept of varying shear rates and shear stresses within a flowing fluid may cause some confusion. First, consider the concept of viscosity. Generally, viscosity gives a measure of fluid thickness. Quantitatively, viscosity

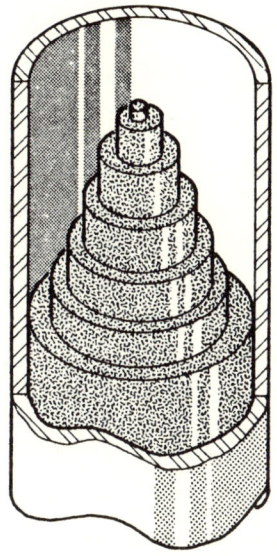

Figure 2-3. Three-Dimensional View of Laminar Flow in a Pipe for a Newtonian Fluid

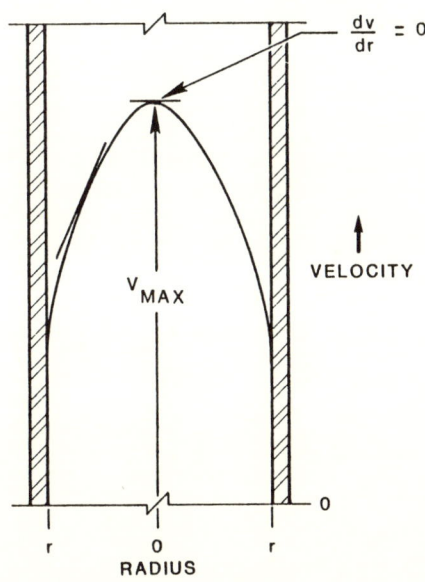

Figure 2-4. Two-Dimensional Velocity Profile of Laminar Flow in a Pipe, for a Newtonian Fluid

is expressed as the ratio of the shear stress to the shear rate. For a Newtonian fluid the ratio is constant and the viscosity is the same at every point throughout the fluid. However, for a non-Newtonian fluid, the viscosity varies with shear rate. Thus for a flowing non-Newtonian fluid, the viscosity varies from one point to another within the pipe. Commonly, one considers viscosity to be a fixed homogeneous characteristic of a substance, but this notion is not accurate for a flowing non-Newtonian fluid.

Second, consider the explanation of shear stress given under paragraph heading 2.2. Under equilibrium conditions, frictional forces are equal in magnitude to the applied force. If shear stress varies with radial distance, then the applied force also must vary with radial distance. Generally, because one usually thinks in terms of a pressure applied across the end of a column of fluid, causing flow, one may erroneously conclude that the pressure also varies with radial distance. The pressure does not vary from point to point across some cross-sectional area of the pipe. From the definition of pressure

$$P = \frac{F}{A}$$

or

$$P * A = F$$

one can see that, at equal pressures, the cylinder shells near the wall, having a greater area than shells near the center, will experience a greater force. In summary, pressure is distributed equally over an end of a column of fluid, but the applied force and shear stress at any cylindrical shell is proportional to the radial distance.

Third, one may wonder how the flow behavior of a fluid can be described if the shear rate and shear stress vary from point to point in the conductor. The practical solution is to measure shear stresses and shear rates at only one specific point in the fluid. Most viscometric instruments measure the shear stress at the conductor wall. Shear stresses are measured at various controlled shear rates, and the fluid behavior thereby may be defined. Theoretically, measuring shear rates and shear stresses at various points in the fluid at a constant average velocity would also define the fluid flow behavior, but such a technique is not practical.

2.6 Turbulent Flow

The velocity profile of a Newtonian fluid in turbulent flow is illustrated in Figure 2-5. Due to the chaotic, random shearing motion, the fluid moves essentially as a plug. Only very near the walls does a thin layer of orderly shear exist. Thus the velocity gradient is very steep near the walls but essentially flat elsewhere.

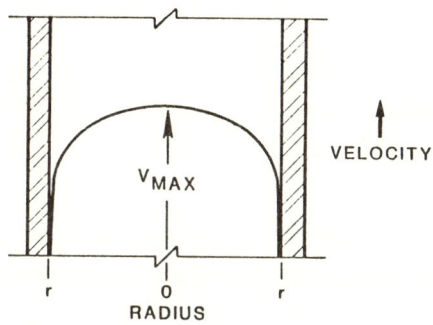

Figure 2-5. Two-Dimensional Velocity Profile of Turbulent Flow in a Pipe, for a Newtonian Fluid

Because very little orderly shear exists in turbulent flow, devices (such as the pipe viscometer and V-G meter) which measure fluid flow properties are designed to evaluate fluids in laminar flow only.

2.7 Regime Determination

As indicated, fluid flow may be either laminar or turbulent. Additionally, there is a region of transitional flow where the the flow is neither completely laminar nor turbulent. The determination of flow regime may be accomplished by calculating either the Reynolds number or the critical velocity. The Reynolds number takes into consideration the basic factors of pipe flow: pipe diameter, average fluid velocity, fluid density, and fluid viscosity. Observing the flow of water in a circular pipe, Reynolds found that turbulence occurred at a value of approximately 2000 when the Reynolds number was defined as

$$Re = \frac{D\bar{v}\rho}{\mu}$$

where

Re = Reynolds number
D = pipe diameter
$\bar{v}$ = average fluid velocity
ρ = fluid density
μ = fluid viscosity

The critical velocity may be found by setting the Reynolds number equal to 2000 and solving the equation for the average fluid velocity.

2.8 CONTINUITY OF FLOW

Consider the flow of a liquid through the pipe at a constant flowrate, Q, as illustrated in Figure 2-6. Because a liquid is nearly incompressible, the volumetric rate of fluid entering the pipe must equal the volumetric rate of fluid leaving the pipe. Thus

$$Q = A_1\bar{v}_1 = A_2\bar{v}_2$$

where

Q = volumetric flowrate
A_1 = area of entrance A_1
A_2 = area of exit A_2
$\bar{v}_1$ = average fluid velocity upon entering A_1
$\bar{v}_2$ = average fluid velocity upon exiting A_2

This is the principle of continuity of flow. The important result of this principle is that, at a constant flowrate, the fluid velocity is inversely proportional to the area through which it flows. The principle may be illustrated by the simple example of placing one's thumb over the end of a water

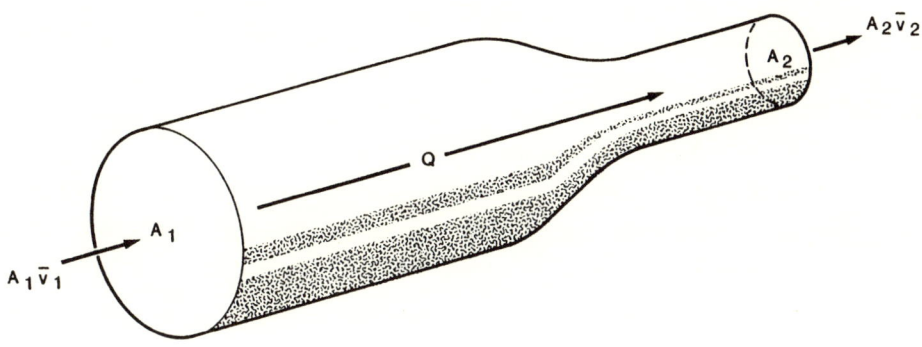

Figure 2-6. Continuity of Flow: Fluid Velocity is Inversely Proportional to the Cross-Sectional Area of the Fluid Conductor

hose to increase the fluid velocity. The thumb decreases the area through which the fluid exits the hose.

Many hydraulic calculations in this manual require the use of the average fluid velocity $\bar{v}$. This quantity may be found by the simple equation

$$\bar{v} = \frac{Q}{A} \tag{2-1}$$

where

$\bar{v}$ = average fluid velocity
Q = volumetric flow rate
A = area of fluid flow

Generally, at a constant flowrate (Q), an average fluid velocity is found for each section of the drillstring and annulus.

2.9 FLUID MODELS

2.10 INTRODUCTION

A fluid model describes the flow behavior of a fluid by expressing a mathematical relationship between shear rate and shear stress. For a Newtonian fluid the ratio of the shear stress to the shear rate is a constant. For non-Newtonian fluids, however, the relationship between shear stress and shear rate is more complex. A generalized relationship for all non-Newtonian fluids has not been found. Instead, various models have been proposed to describe the behavior of several ideal non-Newtonian fluids.

Drilling fluids usually behave as non-Newtonian fluids. Five non-Newtonian models are discussed in this manual: the Bingham Plastic model, the Power Law model, the Casson model, the Robertson-Stiff model, and the Herschel-Bulkley model.

2.11 Newtonian Fluid Model

The Newtonian fluid model is defined by the relationship

$$\tau = \mu \gamma \qquad (2-2)$$

where

τ = shear stress
μ = absolute viscosity
γ = shear rate

At a constant temperature and pressure, the shear rate and shear stress are directly proportional; the constant of proportionality, μ, is the absolute viscosity. Figure 2-7 illustrates the flow curve of a Newtonian fluid. Note that the flow curve is a straight line which passes through the origin, and that the slope of the line is μ, the absolute viscosity.

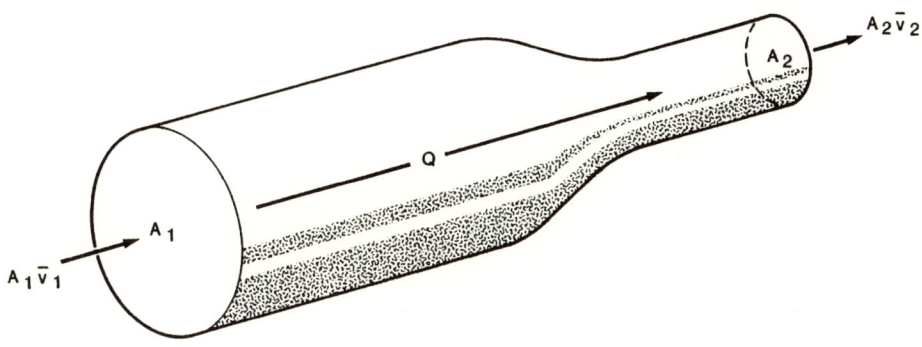

Figure 2-6. Continuity of Flow: Fluid Velocity is Inversely Proportional to the Cross-Sectional Area of the Fluid Conductor

hose to increase the fluid velocity. The thumb decreases the area through which the fluid exits the hose.

Many hydraulic calculations in this manual require the use of the average fluid velocity $\bar{v}$. This quantity may be found by the simple equation

$$\bar{v} = \frac{Q}{A} \tag{2-1}$$

where

$\bar{v}$ = average fluid velocity
Q = volumetric flow rate
A = area of fluid flow

Generally, at a constant flowrate (Q), an average fluid velocity is found for each section of the drillstring and annulus.

2.9 FLUID MODELS

2.10 INTRODUCTION

A fluid model describes the flow behavior of a fluid by expressing a mathematical relationship between shear rate and shear stress. For a Newtonian fluid the ratio of the shear stress to the shear rate is a constant. For non-Newtonian fluids, however, the relationship between shear stress and shear rate is more complex. A generalized relationship for all non-Newtonian fluids has not been found. Instead, various models have been proposed to describe the behavior of several ideal non-Newtonian fluids.

Drilling fluids usually behave as non-Newtonian fluids. Five non-Newtonian models are discussed in this manual: the Bingham Plastic model, the Power Law model, the Casson model, the Robertson-Stiff model, and the Herschel-Bulkley model.

2.11 Newtonian Fluid Model

The Newtonian fluid model is defined by the relationship

$$\tau = \mu\gamma \qquad (2-2)$$

where

τ = shear stress
μ = absolute viscosity
γ = shear rate

At a constant temperature and pressure, the shear rate and shear stress are directly proportional; the constant of proportionality, μ, is the absolute viscosity. Figure 2-7 illustrates the flow curve of a Newtonian fluid. Note that the flow curve is a straight line which passes through the origin, and that the slope of the line is μ, the absolute viscosity.

Water and several pure organic liquids are Newtonian fluids. Drilling fluids rarely behave as Newtonian fluids. Because Newtonian fluids behave in a relatively simple manner, they provide an ideal fluid for fluid flow experiments.

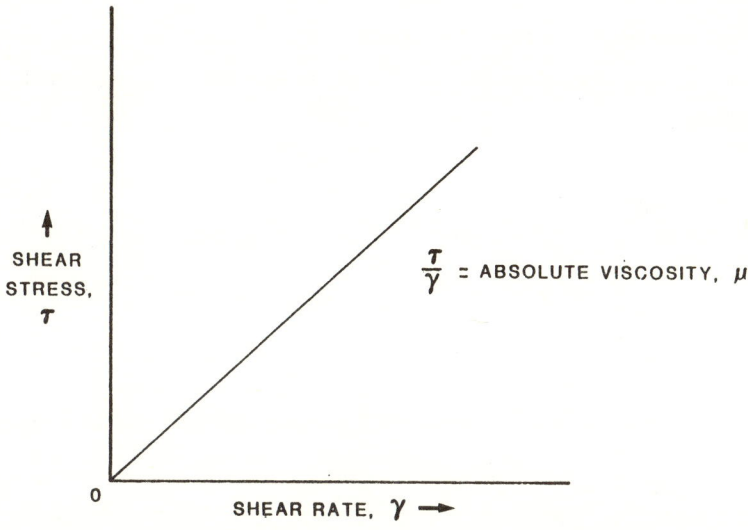

Figure 2-7. Flow Curve for a Newtonian Fluid

2.12 Bingham Plastic Model

The Bingham Plastic fluid model, illustrated in Figure 2-8, is defined by the relationship

$$\tau = \tau_0 + \mu_\infty \gamma \qquad (2-3)$$

where

τ = shear stress
τ_0 = yield stress
μ_∞ = plastic viscosity
γ = shear rate

The Bingham fluid differs most notably from a Newtonian fluid by the presence of a yield stress, commonly called the "yield point." The yield stress is a measure of electrical attractive forces in the mud under flowing conditions. No bulk movement of the fluid occurs until the applied stress exceeds the yield stress. Once the yield stress is exceeded, equal increments of shear stress produce equal increments of shear rate.

The apparent viscosity, or effective viscosity, defined as the shear stress divided by the shear rate, varies with shear rate for non-Newtonian fluids. The apparent viscosity is the slope of a line from the origin to the shear stress at some particular shear rate. The slopes of the dashed lines in Figure 2-8 represent apparent viscosities at various shear rates. The apparent viscosity decreases with increased shear rate. This phenomenon is called "shear thinning."

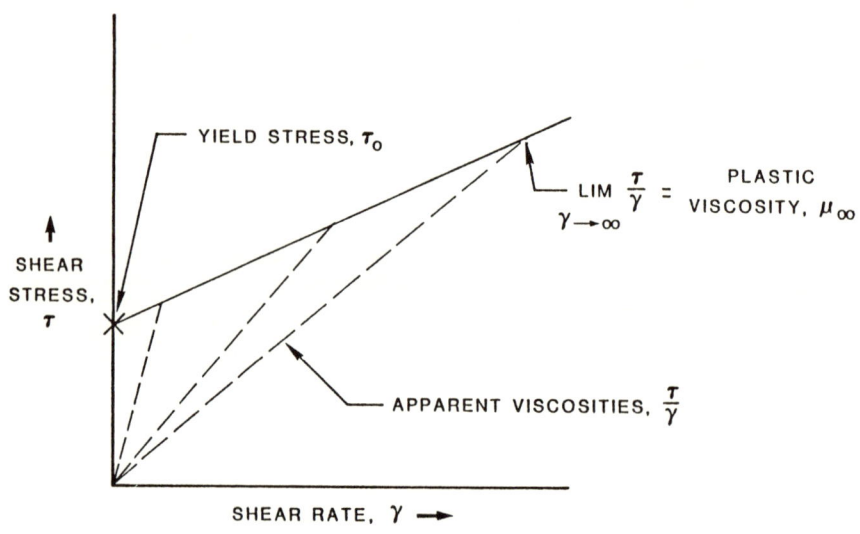

Figure 2-8. Flow Curve for a Bingham Plastic Fluid

As shear rates approach infinity, the apparent viscosity reaches a limit called the "plastic viscosity." The plastic viscosity is the slope of the Bingham Plastic line.

The Bingham Plastic fluid model has been used extensively in the oil industry. The model is easy to use, and it represents many drilling fluids reasonably accurately. In fact, the commonly-used V-G meter was specifically designed to facilitate the use of the Bingham Plastic fluid model in the field. However, the Bingham model usually does not accurately represent drilling fluids at low shear rates, and to some extent more sophisticated fluid models have now replaced the Bingham model in the oilfield.

2.13 Power Law Fluid Model

The Power Law fluid model is defined by the relationship

$$\tau = k\gamma^n \qquad (2\text{-}4)$$

where
- τ = shear stress
- k = consistency factor
- γ = shear rate
- n = flow behavior index

Generally, the consistency factor, k, describes the thickness of the fluid and is thus somewhat analogous to apparent viscosity. As k increases, the mud becomes thicker.

The flow behavior index, n, indicates the degree of non-Newtonian behavior. When $n = 1$, the Power Law equation is identical to the Newtonian fluid equation. If n is greater than 1, the fluid is classified as dilatant; the apparent viscosity increases as shear rate increases. If n is between zero and 1, the fluid is classified as pseudoplastic. Pseudoplastic fluids exhibit shear thinning; that is, the apparent viscosity decreases as the shear rate increases.

For drilling purposes, shear thinning is a very desirable property, and most drilling fluids are pseudoplastic. The two classes of Power Law fluids are illustrated in Figure 2-9.

The Power Law model is used widely in the oil industry and has superseded, to some extent, the Bingham model. The Power Law model is frequently more convenient than the Bingham model: it is particularly suitable for graphical techniques since rotary viscometer readings versus rpm and flow-pressure loss versus flowrate can be plotted as straight lines on log-log paper. The Power Law model also more accurately demonstrates the behavior of a drilling fluid at low shear rates; however, the Power Law model does not include a yield stress and therefore can give poor results at extremely low shear rates.

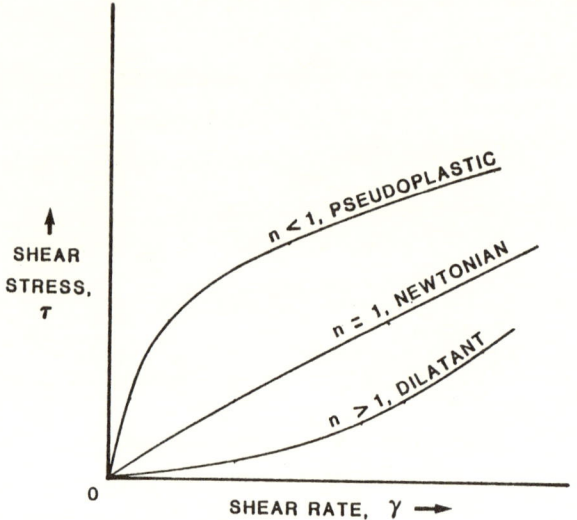

Figure 2-9. Flow Curves for Power Law Fluid

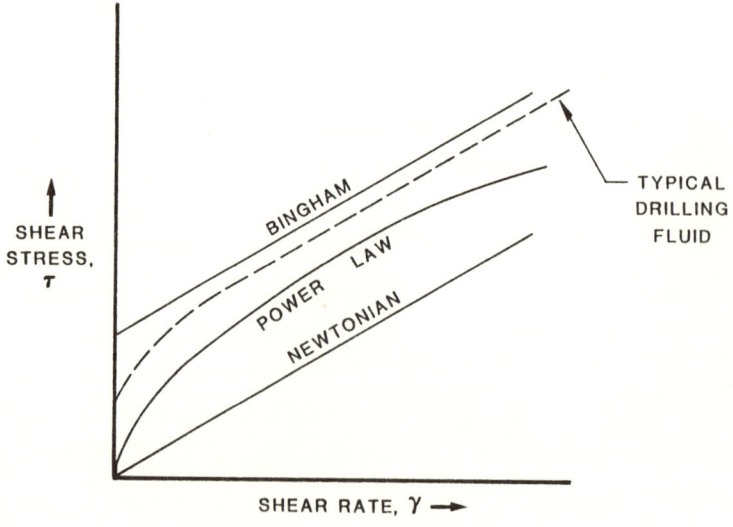

Figure 2-10. Typical Drilling Fluid Vs. Newtonian, Bingham and Power Law Fluids

Figure 2-10 compares the flow curve of a typical drilling fluid to the flow curves of Newtonian, Bingham, and Power Law fluids. A typical drilling fluid exhibits a yield stress and shear thinning. At high rates of shear, all models represent a typical drilling fluid reasonably well. Differences between fluid models are most pronounced at low rates of shear. The Bingham fluid includes a simple yield stress, but does not accurately describe the fluid behavior at low shear rates. The Power Law more accurately describes the behavior at low shear rates, but does not include a yield stress. A typical drilling fluid exhibits behavior intermediate between the Bingham model and the Power Law model.

2.14 Other Models

Three other major models have been developed which at low shear rates exhibit behavior intermediate between that of the Bingham and Power Law models. These are the Casson model, the Robertson-Stiff model, and the Herschel-Bulkley model. The models are defined by

$$\tau = [\tau_0^{.5} + (\mu_\infty \gamma)^{.5}]^2 \quad \text{(Casson)} \tag{2-5}$$

$$\tau = k(\gamma_0 + \gamma)^n \quad \text{(Robertson-Stiff)} \tag{2-6}$$

$$\tau = \tau_0 + k\gamma^n \quad \text{(Herschel-Bulkley)} \tag{2-7}$$

where

τ = shear stress
τ_0 = yield stress
μ_∞ = plastic viscosity
γ = shear rate
k = consistency factor
γ_0 = shear rate intercept
n = flow behavior index

The models are, in effect, hybrid models of the Bingham and Power Law models. The Casson model is a two-parameter model that is used widely in some industries but rarely applied to drilling fluids. The point at which the Casson curve intercepts the shear stress axis varies with the ratio of the yield point to the plastic viscosity. The Robertson-Stiff and Herschel-Bulkley models are three-parameter models which, if necessary, can emulate Bingham or Power Law behavior. The Robertson-Stiff model includes the gel strength as a parameter, and the model is used to a limited extent in the oil industry. The Herschel-Bulkey model is a Power Law model with a yield stress. Unfortunately, for most oilfield applications, the Herschel-Bulkley model yields mathematical expressions which are not readily solvable. Therefore, for practical reasons the Herschel-Bulkley model is not used in the oil industry. However, some drilling fluid companies have made use of a "pseudo" Herschel-Bulkley model. This pseudo model, however, does not allow for a direct correlation between viscometric analysis and hydraulic calculations.

2.15 TIME-DEPENDENT BEHAVIOR

Non-Newtonian fluids may exhibit time-dependent flow behavior: the apparent viscosity at a fixed shear rate does not remain constant, but varies to some maximum or minimum with the duration of shear. If the apparent viscosity _decreases_ with flow time, the fluid is thixotropic. Conversely, once flow has ceased, a thoxitropic fluid will exhibit increasing apparent viscosity with increasing time. If the apparent viscosity _increases_ with flow time, the fluid is rheopectic.

Most drilling fluids exhibit time-dependent flow behavior. The shear stress developed in most drilling fluids is dependent on the duration of shear. A time lag exists between an adjustment of the shear rate and stabilization of the shear stress at its corresponding value. This occurs primarily because the clay plates or fibers are broken into smaller particles at higher rates of shear, and aggregate into layer units as shear rate is decreased again, with both of these events taking a noticeable length of time. In practice, one finds that the shear stress stabilizes much more quickly when shear rate is reduced rather than increased. Thus, in a rotating viscometer, drilling fluid should be sheared at a high speed for some time before its properties are measured.

Gel strength measurements describe the time-dependent flow behavior of a drilling fluid. Whereas the yield stress measures the attractive forces of the fluid under flowing conditions, the gel strength measures the attractive forces of a drilling fluid under static conditions. Gel strength increases with time. If the gel strength increases steadily with time, the gel strength is classified as strong, or "progressive." If the gel strength increases only slightly with time, the gel strength is classified as weak or "fragile." Strong and weak gel strengths are illustrated in Figure 2-11.

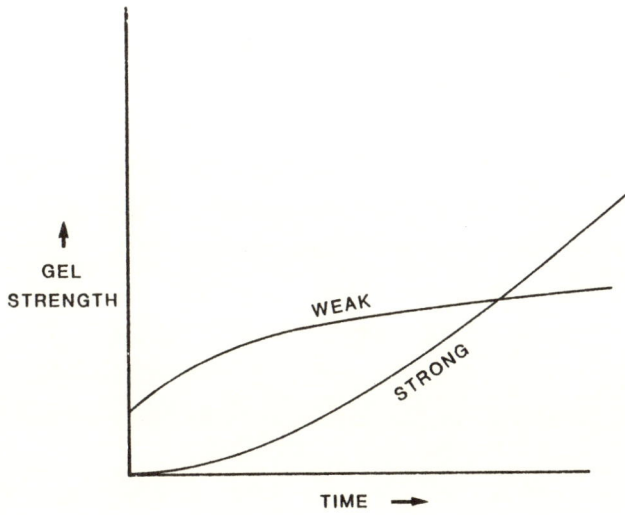

Figure 2-11. Gel Strengths

Strong gels are usually caused by a high concentration of clays. Problems such as excessive pressure to break circulation and lost circulation may result from strong gels. Thus, strong gels are undesirable.

2.16 MEASUREMENT OF FLUID FLOW PROPERTIES

2.17 INTRODUCTION

The measurement of fluid flow properties represents a transitional stage in the discussion of drilling hydraulics. Up to this point the discussion has been concerned primarily with the basic principles of fluid hydraulics. A discussion of fluid mechanics that is more specific to drilling hydraulics now begins. Such topics as viscometric measurements, pressure losses, swab and surge pressures, and bit hydraulics optimization are included.

The measurement of fluid flow properties may be accomplished with a variety of viscometric instruments. The following text rigorously analyzes the measurement of fluid flow properties with a rotating-sleeve viscometer only, the most commonly used viscometric instrument.

The analysis is fairly complex. It involves the mathematical analysis of shear rates and shear stresses within the conductor through which the fluid flows. The mathematical analysis is included to make this manual a comprehensive text. However, the mathematical discussion may be bypassed without loss of understanding of the subsequent material.

2.18 INSTRUMENTS

Three instruments are commonly used to measure viscometric properties: the pipe flow rheometer, the Marsh funnel, and the rotating-sleeve viscometer. The pipe flow rheometer is a sophisticated instrument that is more often found in the laboratory than in the oilfield. Fluid is pumped through a horizontal pipe at a measured rate, and fluid pressure is measured at two points along the pipe. The difference between these pressure readings is a measure of shear stress at the pipe wall, while shear rate at the wall is a function of flowrate and fluid properties. Flow of fluid through pipes will be covered in Section 5, so interpretation of pipe flow data will not be discussed here.

Two instruments are widely used in the oilfield to measure fluid flow properties -- the Marsh funnel and the rotating-sleeve viscometer.

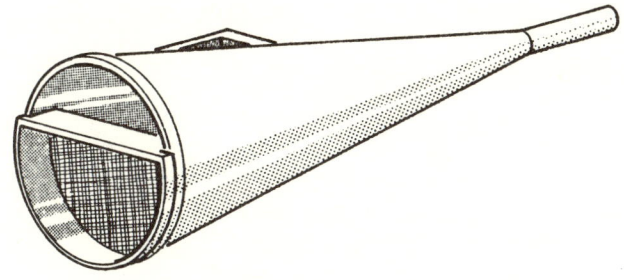

Figure 2-12. Marsh Funnel

The Marsh funnel (illustrated in Figure 2-12) consists of a conical funnel with a short, narrow tube attached. The funnel is filled to the mark, and the time required for one U.S. quart of fluid to drain through the tube is then measured. Marsh-funnel viscosity is measured in seconds, and typically may range from 30 to 80 seconds. Pure water has a funnel viscosity of 26 seconds. The Marsh-funnel viscosity may also be defined as the time taken to drain one liter, rather than one quart (946 cc).

Since only one measurement is made with the Marsh funnel, it cannot be used to determine the viscous properties of a non-Newtonian fluid. The funnel viscosity is a function of fluid density as well as viscosity, and the reading should therefore be used only in a qualitative sense. The Marsh funnel is, however, a rugged field tool and can indicate gross changes in fluid behavior. When such changes are noted, more precise measurements are required.

Such measurements are usually obtained with a rotating-sleeve viscometer, as illustrated in Figure 2-13. The outer sleeve is rotated at a known speed and torque is transmitted through the fluid to the cylindrical bob. Many field models are hand-operated and offer only two speeds (usually 300 and 600 rpm), but electric viscometers may offer a wide range of speeds. The bob is connected to a spring and dial, where the torque can be measured. It is also possible to use springs and bobs of various dimensions to investigate different shear-rate ranges. The figures read from the dial are degrees of rotation; and a conversion factor, the value of which depends on the spring and bob combination and on the units required, is needed to convert the dial reading to units of shear stress.

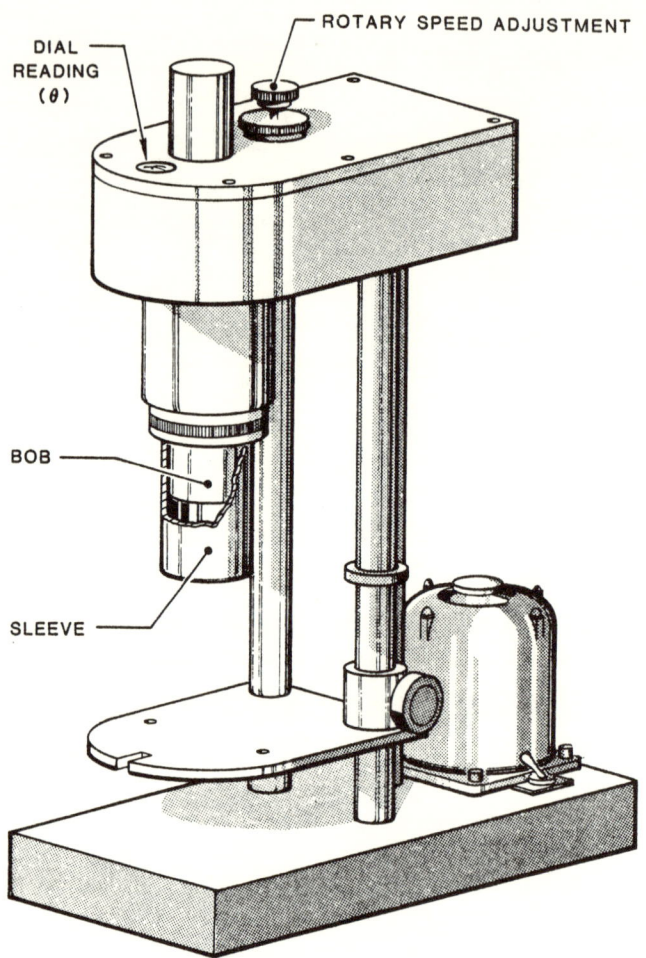

Figure 2-13. Rotating-Sleeve Viscometer

2.19 ANALYSIS

The analysis of fluid flow properties with the rotating-sleeve viscometer depends upon several assumptions:

- The fluid is in laminar flow.
- The fluid system is in equilibrium.
- The fluid velocity at the bob is zero.
- The fluid is completely defined by the fluid model.
- The fluid behavior is not time-dependent.

The analysis of shear stresses and shear rates in a fluid depends upon orderly shear, or laminar flow. The rotating-sleeve viscometer is designed to ensure laminar flow between the sleeve and the bob for all typical oilfield viscosities and shear rates.

In the equilibrium state, no net forces act upon the fluid. This situation was described in detail under paragraph heading 2.2. The assumption of an equilibrium state allows the shear stress to be defined as a function of radial distance.

The assumption of zero fluid velocity at the bob is a specific example of the fundamental assumption of fluid hydraulics which states that a fluid in contact with a surface has the same velocity as that surface. The velocity of the fluid at the sleeve is equal to the viscometer rotary speed at the sleeve and is zero at the stationary bob.

The assumption that the fluid is completely defined by the fluid model enables the shear rate to be expressed as a function of shear stress; thus, the shear rate may be expressed as a function of radial distance as well.

A non-time-dependent fluid behavior must be assumed in order to derive a meaningful analysis of the fluid behavior. Any changes with time must be ignored by the analysis.

The general logic behind the analysis is to obtain a desired result in terms of what is already known. The goal of the analysis is to completely define the flow behavior at any point in the fluid in terms of shear rate and shear stress. From the assumptions, much is already known: the shear rate and shear stress are some function of radial distance, the fluid velocity at the bob is zero. At the sleeve, the fluid velocity is controllable. In the equilibrium state, the force applied to the fluid is equal to the friction forces which are measurable as a shear stress at the bob. By finding an expression for shear rate and shear stress at the bob in particular, several readings may be taken and the rheological properties of the fluid may be established for each fluid model.

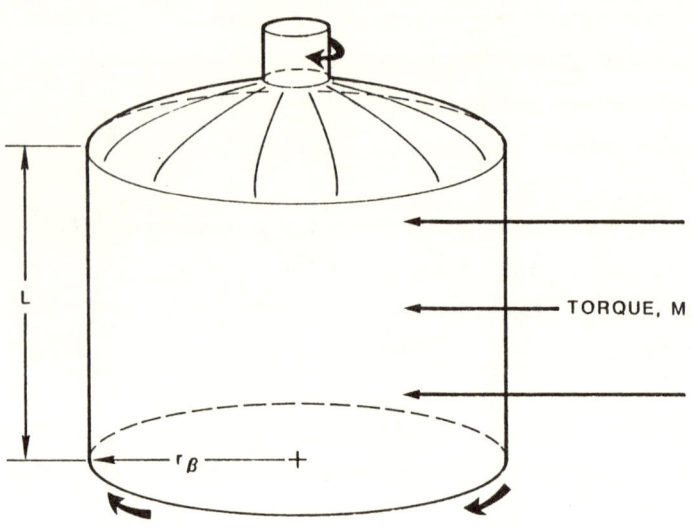

Figure 2-14. Torque Acting on the Bob

Torque is a force that causes an object to rotate or twist rather than to linearly accelerate. The torque applied at the bob is equal to the shear stress at the bob, multiplied by the area over which it acts, multiplied by the radius. Effects upon the bottom surface of the bob are small and may be ignored. Thus from Figure 2-14,

$$\tau_\beta = \frac{M}{2\pi r_\beta^2 L} \qquad (2-8)$$

where

τ_β = shear stress at the bob
M = moment, or torque
r_β = radius of bob
L = length

The fluid between the bob and the sleeve may be viewed as innumerable thin cylindrical shells that are sliding past one another. Because the system is in equilibrium, no net torque acts upon the system. The torque that is sliding one shell past another is equal to the frictional forces between the shells.

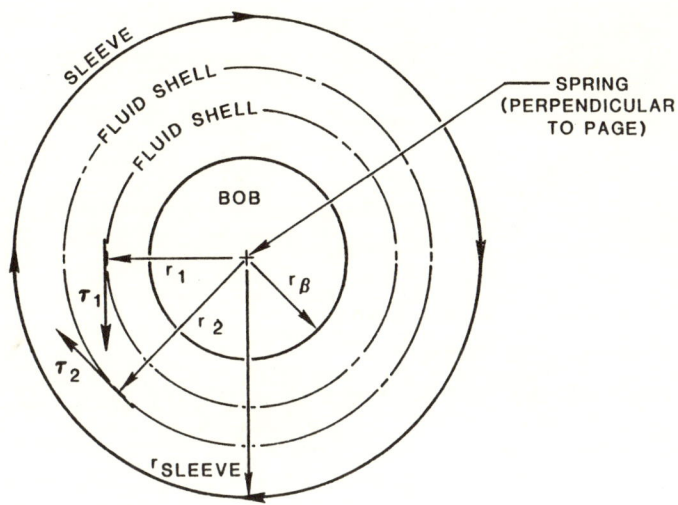

Figure 2-15. Forces in Equilibrium on Rotating-Sleeve Viscometer

Thus from Figure 2-15,

$$M_2 = \tau_2 * r_2 * 2\pi r_2 * L$$

$$M_1 = \tau_1 * r_1 * 2\pi r_1 * L$$

where

    M = the torque acting upon the shells
    τ = shear stress
    r = shell radius
    L = length

and

$$M_2 = M_1$$

$$\tau_2 * r_2 * 2\pi r_2 * L = \tau_1 * r_1 * 2\pi r_1 * L$$

$$\tau_2 = \tau_1 * \frac{r_1^2}{r_2^2}$$

Thus, the shear stress at any point in the fluid is inversely proportional to the square of the radial position. This may be stated also as

$$\tau = \tau_\beta * \frac{1}{y^2} \qquad (2\text{-}9)$$

where

τ = shear stress at any point
τ_β = shear stress at the bob
y = ratio of the radius at any point to the radius of the bob, or $\frac{r}{r_\beta}$

Expressing the fluid model in general as $\gamma = f(\tau)$, this gives

$$\gamma = f\left(\tau_\beta * \frac{1}{y^2}\right) \qquad (2\text{-}10)$$

Finally, the angular velocity at any point is expressed as a function of shear rate. Figure 2-16 illustrates the displacement between two fluid shells moving at different angular velocities:

$$\frac{dv}{dr} = \frac{v_2 - v_1}{r_2 - r_1} = \frac{\omega_2 r_2 - \omega_1 r_1}{r_2 - r_1}$$

If $\omega_2 = \omega_1 + d\omega$ and $r_2 = r_1 + dr$, then

$$\frac{dv}{dr} = \frac{(\omega_1 + d\omega)(r_1 + dr) - \omega_1 r_1}{(r_1 + dr) - r_1}$$

which simplifies to

$$\frac{dv}{dr} = r_1 \frac{d\omega}{dr} + \omega_1 + d\omega \qquad (2\text{-}11)$$

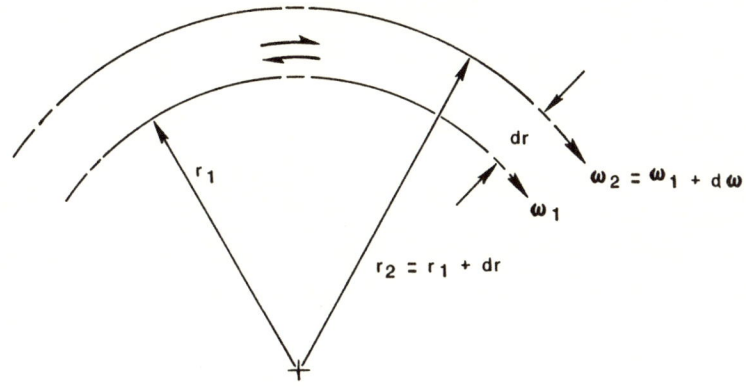

Figure 2-16. Displacement Between Two Fluid Shells Moving at Different Angular Velocities

Equation (2-11) expresses the total change of velocity with radius. The last two terms of this equation represent the change of velocity that results even with no shear. Thus, the change in velocity with radial position due to shear, or the shear rate, is given by

$$\gamma = \frac{dv}{dr} = r_1 \frac{d\omega}{dr} \qquad (2\text{-}12)$$

Because y was defined in Equation (2-9) as the ratio of the radius at any point to the radius of the bob, Equation (2-12) may be rewritten as

$$\gamma = \frac{dv}{dr} = y * \frac{d\omega}{dy} \qquad (2\text{-}13)$$

Equations (2-10) and (2-13) are both expressions of shear rate and therefore are equal:

$$f\left(\tau_\beta * \frac{1}{y^2}\right) = y * \frac{1}{dy} * d\omega$$

or,

$$d\omega = f\left(\tau_\beta * \frac{1}{y^2}\right) * \frac{1}{y} * dy$$

And upon integration and rearrangement,

$$\omega = \int \frac{1}{y} * f\left(\tau_\beta * \frac{1}{y^2}\right) * dy \qquad (2\text{-}14)$$

Angular velocity at the sleeve is determined by the rotating speed of the viscometer, given in units of rpm. In radian units, the rpm speed becomes

$$N * rpm = N * \frac{revolution}{minute}$$

$$= N * \frac{2\pi}{minute} * \frac{1\ minute}{60\ seconds}$$

$$= \frac{2\pi N}{60}$$

Thus, because the angular velocity is zero at the bob and $2\pi N/60$ at the sleeve, Equation (2-14) becomes

$$\frac{2\pi N}{60} = \int_1^b \frac{1}{y} * f\left(\tau_\beta * \frac{1}{y^2}\right) * dy \qquad (2\text{-}15)$$

where

    b = radius of sleeve/radius of bob

The integration is carried out from the bob, where y = radius of the bob/radius of the bob = 1, to the sleeve, where y = b = radius of the sleeve/radius of the bob.

Equation (2-15) can be used to find the shear stress at the bob for a given rpm for each fluid model. For example, for a Newtonian fluid

$$\gamma = f(\tau) = \frac{\tau}{\mu}$$

$$\frac{2\pi N}{60} = \int_1^b \frac{1}{y} * \frac{\tau_\beta}{y^2 \mu} * dy$$

$$\frac{2\pi N}{60} = \int_1^b \frac{\tau_\beta}{y^3 \mu} * dy$$

$$\frac{2\pi N}{60} = \frac{\tau_\beta}{\mu} \int_1^b \frac{1}{y^3} * dy$$

$$\frac{2\pi N}{60} = \frac{\tau_\beta}{\mu} \left[-\frac{1}{2y^2} \right]_1^b$$

$$\frac{2\pi N}{60} = \frac{\tau_\beta}{2\mu} \left[-\frac{1}{b^2} + 1 \right]$$

$$\frac{2\pi N}{60} = \frac{\tau_\beta}{2\mu} \left(\frac{b^2 - 1}{b^2} \right)$$

$$\tau_\beta = 2\mu \left(\frac{b^2}{b^2 - 1} \right) \left(\frac{2\pi N}{60} \right)$$

$$\tau_\beta = \frac{\pi N b^2 \mu}{15(b^2 - 1)}$$

For the commonly-used fluid models, the shear stress at the bob is expressed as follows:

Newtonian:
$$\tau_\beta = \frac{\pi N b^2 \mu}{15(b^2 - 1)} \tag{2-16}$$

Bingham:
$$\tau_\beta = \frac{2b^2}{(b^2 - 1)} * \left[\frac{2\pi N \mu_\infty}{60} + \tau_0 \ln b\right] \tag{2-17}$$

Power Law:
$$\tau_\beta = k \left[\frac{\pi N * b^{2/n}}{15n (b^{2/n} - 1)}\right]^n \tag{2-18}$$

Casson: (2-19)

$$\tau_\beta = \left[\frac{2b(b-1) * \tau_0^{.5} + \sqrt{4b^2(b-1)^2 * \tau_0 + 2b^2(b^2-1) * \left(\frac{\pi N \mu_\infty}{30} - \tau_0 \ln b\right)}}{(b^2 - 1)}\right]^2$$

Robertson-Stiff:
$$\tau_\beta = k \left[\frac{2 * b^{2/n}}{n(b^{2/n}-1)} * \left(\frac{2\pi N}{60} + \gamma_0 \ln b\right)\right]^n \tag{2-20}$$

The fluid model may be used to convert shear stress at the bob to shear rate at the bob, γ_β. For example, for a Newtonian fluid,

$$\gamma = f(\tau) = \frac{\tau}{\mu}$$

$$\gamma_\beta = \frac{\tau_\beta}{\mu}$$

Substituting Equation (2-16),

$$\gamma_\beta = \frac{\left[\frac{\pi N b^2 \mu}{15(b^2-1)}\right]}{\mu} = \frac{\pi N b^2}{15(b^2-1)}$$

For the commonly-used fluid models, shear rate at the bob is expressed as follows:

Newtonian:

$$\gamma_\beta = \frac{\pi N b^2}{15(b^2-1)} \qquad (2\text{-}21)$$

Bingham:

$$\gamma_\beta = \frac{\pi N b^2}{15(b^2-1)} + \frac{\tau_0}{\mu_\infty} * \left[\frac{2b^2 \ln b}{(b^2-1)} - 1\right] \qquad (2\text{-}22)$$

Power Law:

$$\gamma_\beta = \frac{\pi N b^{2/n}}{15n(b^{2/n}-1)} \qquad (2\text{-}23)$$

Casson:

$$\gamma_\beta = \left[\frac{(b-1)^2 \tau_0^{.5} + \sqrt{4b^2(b-1)^2 \tau_0 + 2b^2(b^2-1)*\left(\frac{\pi N \mu_\infty}{30} - \tau_0 \ln b\right)}}{(b^2-1)\mu_\infty^{.5}}\right]^2 \qquad (2\text{-}24)$$

Robertson-Stiff:

$$\gamma_\beta = \frac{\pi N b^{2/n}}{15n(b^{2/n}-1)} + \gamma_0 \left[\frac{2b^{2/n} \ln b}{n(b^{2/n}-1)} - 1\right] \qquad (2-25)$$

The shear stress and shear rate at the bob cannot be expressed analytically for a Herschel-Bulkley fluid. If fluid properties are measured with the rotating-sleeve viscometer, this gives the Robertson-Stiff model a marked advantage over the Herschel-Bulkley model. It will be shown in this manual that the Robertson-Stiff model offers similar advantages in the analysis of pipe flow and annular flow.

2.20 RESULTS

Since we can now express shear rate at the bob in terms of rpm and the fluid properties, and since the shear stress can be measured at a number of different speeds, it is possible to determine the fluid properties if the number of measurements equals the number of parameters in the fluid model.

For a Newtonian fluid:

$$\mu = \frac{15 \tau_\beta (b^2-1)}{\pi N b^2} \qquad (2-26)$$

For a Bingham fluid:

$$\mu_\infty = \frac{15 (b^2-1)(\tau_2 - \tau_1)}{\pi(N_2 - N_1) b^2} \qquad (2-27)$$

$$\tau_o = \frac{(b^2-1)(N_2\tau_1 - N_1\tau_2)}{2b^2 \ln b (N_2 - N_1)} \qquad (2\text{-}28)$$

For a Power Law fluid:

$$n = \frac{\log(\tau_2/\tau_1)}{\log(N_2/N_1)} \qquad (2\text{-}29)$$

$$k = \tau_1 \left[\frac{15n(b^{2/n}-1)}{\pi N_1 b^{2/n}}\right]^n \qquad (2\text{-}30)$$

For a Casson fluid: (2-31)

$$\tau_o = \left[\frac{4b(b-1)\left(N_2\tau_1^{.5} - N_1\tau_2^{.5}\right)}{} \quad \text{(continued below)}\right.$$

$$\left.\frac{-\sqrt{16b^2(b-1)^2(N_2\tau_1^{.5} - N_1\tau_2^{.5})^2 - 8b^2(b^2-1)\ln b(N_2\tau_1 - N_1\tau_2)(N_2-N_1)}}{4b^2 \ln b (N_2 - N_1)}\right]^2$$

$$\mu_\infty = \frac{30}{\pi N_2} * \left[\frac{\tau_2(b^2-1) - 4b(b-1)(\tau_o\tau_2)^{.5}}{2b^2} + \tau_o \ln b\right] \qquad (2\text{-}32)$$

For a Robertson-Stiff fluid:

$$\left(\frac{\tau_1}{\tau_2}\right)^{1/n} * \frac{(N_3 - N_2)}{(N_3 - N_1)} + \left(\frac{\tau_3}{\tau_2}\right)^{1/n} * \frac{(N_2 - N_1)}{(N_3 - N_1)} = 1 \qquad (2\text{-}33)$$

$$\gamma_o = \frac{\pi \left(N_3 \tau_2^{1/n} - N_2 \tau_3^{1/n} \right)}{30 \ln b \left(\tau_3^{1/n} - \tau_2^{1/n} \right)} \qquad (2\text{-}34)$$

$$k = \tau_2 * \left[\frac{n \left(1 - b^{-2/n} \right)}{\frac{2\pi N}{15} + 2\gamma_o \ln b} \right]^n \qquad (2\text{-}35)$$

For the above equations N_3, N_2, and N_1 are rpm speeds where $N_3 > N_2 > N_1$, which give shear stresses τ_3, τ_2, and τ_1, respectively. Equation (2-33) must be solved iteratively to find n.

2.21 FIELD PROCEDURES

While Equations (2-26) through (2-35) may appear unfamiliar and perhaps intimidating, they can usually be simplified considerably when the viscometer dimensions are known. When fitted with the number-one bob, the Fann viscometer (used widely in the oilfield) has a diameter ratio (b) of 1.0678. The number-one spring and number-one bob combination gives a reading of 1.067 lb/100 ft^2 per degree of rotation. For fluids which do not obey the Newtonian or Bingham fluid models, the shear stress at a given rotary speed is the dial reading, θ, multiplied by 1.067. Substituting these values into Equation (2-26) yields

$$\mu = \frac{15 \tau_\beta (b^2 - 1)}{\pi N b^2}$$

$$\mu = \frac{15 (1.067 \theta_1) \left[(1.0678)^2 - 1 \right]}{\pi N_1 (1.0678)^2}$$

$$1.5964 \mu = \frac{\theta_1}{N_1} \qquad (2\text{-}36)$$

where

θ_1 = the dial reading at rotary speed N_1

For a Bingham fluid, Equations (2-27) and (2-28) become

$$1.5964\mu_\infty = \frac{\theta_2 - \theta_1}{N_2 - N_1} \qquad (2\text{-}37)$$

$$\tau_o = \frac{(N_2\theta_1 - N_1\theta_2)}{N_2 - N_1} \qquad (2\text{-}38)$$

If rotation speeds are N_1 = 300 and N_2 = 600 rpm, the constant in Equations (2-36) and (2-37) becomes 478.9, which is exactly the units conversion from lb/sec/100 ft^2 to centipoise. We therefore have the important result that the Fann viscometer with 1/1 spring-and-bob combination, when run at 300 rpm, gives a direct dial reading of Newtonian viscosity in centipoise. If the fluid is of Bingham type, the difference between 600 rpm and 300 rpm readings gives the plastic viscosity in centipoise, while twice the 300 rpm reading, minus the 600 rpm reading, gives the yield point. These results did not come about by accident: the dimensions of the viscometer were designed to permit direct measurement of Newtonian or Bingham fluid properties in the usual oilfield units.

Application of Equations (2-29) through (2-35) to actual oilfield calculations is relatively simple, using the forms of these equations in Appendix C (which include units conversions). The case of a Power Law fluid will be briefly discussed here, because calculation of Power Law parameters is frequently required in the field.

The flow behavior index n can be calculated directly from Equation (2-29). Dial readings may be used directly, without converting to shear stress, since only the ratio of the stresses is used in the calculation. Calculation of the consistency factor k is slightly more complicated. Substituting the above dimensions and conversion factors in Equation (2-30) gives

$$k = 1.067 \ \theta \left[\frac{n}{20\pi} \left(1 - 1.0678^{-2/n}\right)\right]^n \tag{2-39}$$

This gives an exact value of k in lb/sec/100 ft².

In the field, the mud engineer usually does not calculate an exact value for k from Equation (2-39). An apparent value of k is calculated at a rotary speed of 300 rpm by

$$k = \frac{\theta_{300}}{511^n} \tag{2-40}$$

where

θ_{300} = dial reading at 300 rpm

The apparent value is obtained by ignoring the dial factor of 1.067 and assuming a Newtonian shear rate of 511 sec^{-1}. The Newtonian shear rate of 511 sec^{-1}, commonly seen in oilfield hydraulics calculations, is the shear rate at the bob at 300 rpm for a Newtonian fluid and may be obtained by substituting N = 300 into Equation (2-21).

A correction factor, called G_v, may be calculated to convert an apparent value of k to a true value of k:

True k = apparent k * correction factor

$$k = \frac{\theta_{300}}{511^n} * G_v$$

$$G_v = k * \frac{511^n}{\theta_{300}} \tag{2-41}$$

Combining Equations (2-39) and (2-41),

$$G_v = \left(\frac{511^n}{\theta_{300}}\right)\left(1.067\theta_{300}\right)\left[\frac{n}{20\pi}\left(1 - 1.0678^{-2/n}\right)\right]^n$$

which simplifies to

$$G_v = 1.067\left[8.1328n\left(1 - 1.0678^{-2/n}\right)\right]^n \qquad (2\text{-}42)$$

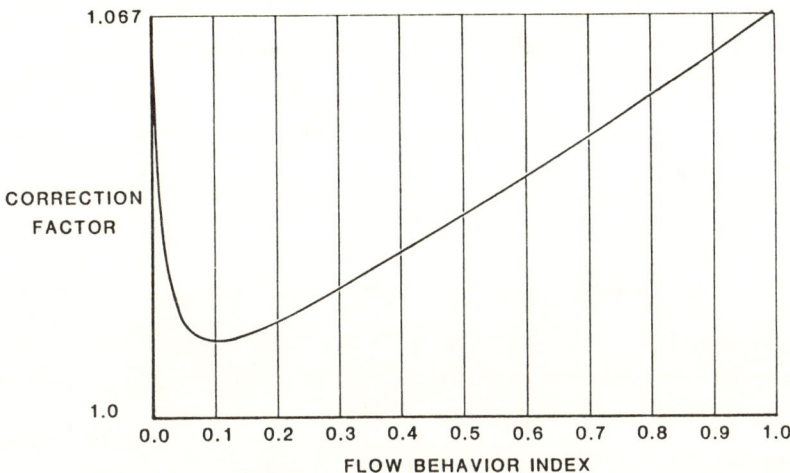

Figure 2-17. K-Correction Factor

This correction factor is graphed in Figure 2-17. There are two common oilfield approximations: one is to ignore the correction factor and to calculate according to Equation (2-40), while the other assumes the correction factor to be equal to the dial factor of 1.067. It can be seen that neither approximation is accurate, but that a very good approximation would be given by = (1 + .067n) for values of n greater than 0.2 (values less than 0.2 are rarely observed). Thus the Power Law consistency factor can be approximated by

$$k \simeq (1 + .067n) * \frac{\theta}{511^n} \qquad [n \geq 0.2] \qquad (2\text{-}43)$$

although the exact form of Equation (2-32) is preferred.

A note of caution is required to complete this analysis of rotary viscometer data interpretation. If the fluid is thought to have a non-zero yield stress, the above methods are valid only if the fluid is in laminar flow throughout the space between sleeve and bob. At very low rates of rotation a plug flow region may develop near the sleeve, although this only occurs when measured shear stress is within the ratio b^2 of yield stress. Thus, there will be no problem if the readings used in calculations are at least a factor b^2 greater than the smallest obtainable reading. In any event, it is unlikely that a large error would result from this source.

In the field, it is often necessary to determine which fluid model best describes the drilling fluid. Most computerized hydraulic programs currently offer only a choice between the Bingham and Power Law models. However, more complex computer programs that allow the choice of any of the five fluid models have been developed by Exlog and are being evaluated in the field.

At the wellsite, the choice of an accurate model is best accomplished by graphing the viscometric data available from the mud engineer. This data typically consists of three quantities: the 600 rpm reading, the 300 rpm reading, and the gel strength. A graph of a typical drilling fluid versus the Newtonian, Bingham, and Power Law fluid models was given in Figure 2-10 and is repeated below in Figure 2-18.

One can see that the position of the gel strength along the shear stress axis predominantly determines which model is the best fit. If the gel strength is high and near the yield point, the fluid is best approximated by the Bingham model. If the gel strength is very low, the fluid is better approximated by the Power Law model.

The Casson model intercepts the shear stress axis at some point between the origin and the yield point that depends upon the ratio of the yield point to the plastic viscosity. A graph of the function is given in Figure 2-19. The intercept on the shear stress axis is given as a percentage of the yield point. By comparing the gel strength to the intercept point, one may determine the accuracy of using the Casson model.

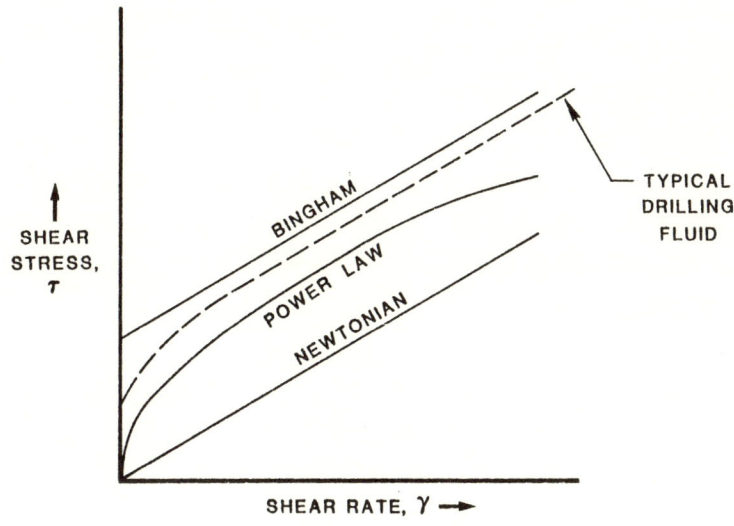

Figure 2-18. Typical Drilling Fluid Vs. Newtonian, Bingham and Power Law Fluids

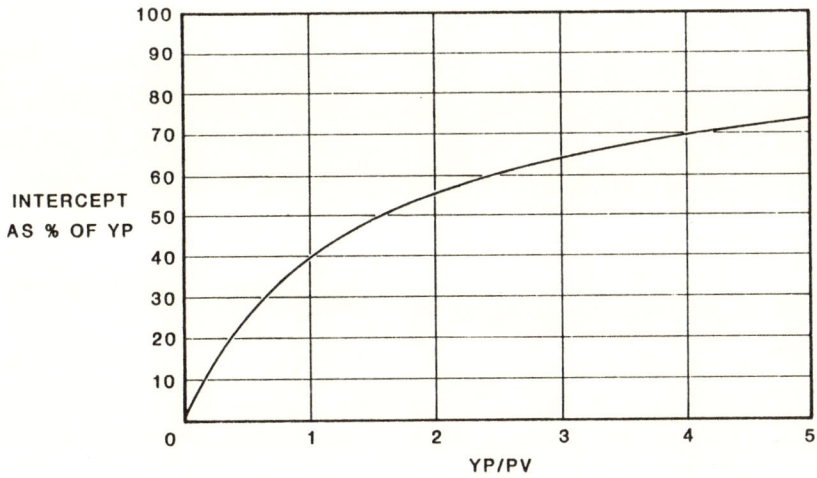

Figure 2-19. Casson Fluid Intercept on the Shear Stress Axis

The Robertson-Stiff model actually includes the gel strength in its curve. Thus, if a sufficiently powerful hydraulic program is available, the Robertson-Stiff model usually will be the most accurate model to use for hydraulics calculations. Note, however, that the Newtonian, Bingham, and Power Law models are all just specific cases of the Robertson-Stiff model, and that in some instances the Robertson-Stiff model would yield to one of these simpler models for hydraulics calculations. The Robertson-Stiff model cannot, however, emulate a Casson fluid.

One last word must be said concerning the determination of a fluid's viscous properties: fluid properties need not remain constant; in fact, viscosity is very dependent upon temperature. The viscous properties determined from mud taken from the flowline can only approximate the viscous properties downhole. High temperature and high pressure rheometers are desirable but as yet are unavailable in the field.

2.22 REFERENCES

1. Bannister, C. E., "Rheological Evaluation of Cement Slurries: Methods and Models," S.P.E. Paper 9284, 1980.

2. Beirute, R. M., and R. W. Flumerfelt, "Mechanics of the Displacement Process of Drilling Muds by Cement Slurries Using an Accurate Rheological Model," S.P.E. Paper 6801, 1977.

3. Beirute, R. M., and R. W. Flumerfelt, "An Evaluation of the Robertson-Stiff Model Describing Rheological Properties of Drilling Fluids and Cement Slurries," S.P.E. J., April 1977.

4. Bootwala, I., "Method Speeds Drill Site Hydraulics Calculation," World Oil, April 1979.

5. Cloud, J. E., and P. E. Clark, "Stimulaton Fluid Rheology III: Alternatives to the Power Law Fluid Model for Cross-Linked Fluids," S.P.E. Paper 9332, 1980.

6. Fann Instrument Corporaton, "Operating Instructions - Fann Viscometers."

7. Fredrickson, A. G., "Helical Flow of an Annular Mass of Visco-Elastic Fluid," Chem. Eng. Sci 11, 1960.

8. Havenaar, I., "The Pumpabilty of Clay-Water Drilling Fluids," S.P.E. Pet. Trans. Reprint Series 6 (Drilling).

9. Howell, J. N., "Improved Method Simplifies Friction-Pressure-Loss Calculations," O&G J., April 25, 1966.

10. Lauzon, R. V., and K.I.G. Reid, "New Rheological Model Offers Field Alternative," O&G J., May 21, 1979.

11. McMordie, W. C., R. B. Bennett and R. G. Bland, "The Effect of Temperature and Pressure on the Viscosity of Oil-Base Muds," J.P.T., July 1975.

12. Melrose, J. C., and W. B. Lilienthal, "Plastic Flow Properties of Drilling Fluids - Measurement and Application," S.P.E. Pet. Trans. Reprint Series 6 (Drilling).

13. Melton, L. L., and W. T. Malone, "Fluid Mechanics Research and Engineering Application in non-Newtonian Fluid Systems, S.P.E. J., March 1964.

14. Robertson, R. E., and H. A. Stiff, "An Improved Mathematical Model for Relating Shear Stress to Shear Rate in Drilling Fluids and Cement Slurries," S.P.E. J., February 1976.

15. Savins, J. G., "Generalized Newtonian (Pseudoplastic) Flow in Stationary Pipes and Annuli," Pet. Trans. A.I.M.E. 213, 1958.

16. Savins, J. G., "The Characterization of non-Newtonian Systems by a Dual Differentiation-Integration Method," S.P.E. J., June 1962.

17. Savins, J. G., and W. F. Roper, "A Direct-Indicating Viscometer for Drilling Fluids," Drilling and Producing Practices, A.P.I., 1954.

18. Savins, J. G., and G. C. Wallick, "Viscosity Profiles, Discharge Rates, Pressures, and Torques for a Rheologically Complex Fluid in a Helical Flow," A.I.Ch.E. J. $\underline{12}$ (2), 1966.

19. Savins, J. G., G. C. Wallick and W. R. Foster, "The Differentiation Method in Rheology - I, Poiseuille Type Flow," S.P.E. J., September 1962.

20. _____ - II, "Characteristic Derivatives of Ideal Models in Poiseuille Flow," S.P.E. J., December 1962.

21. _____ - III, "Couette Flow," S.P.E. J., March 1963.

22. _____ - IV, "Characteristic Derivatives of Ideal Models in Couette Flow," S.P.E. J., June 1963.

23. Taylor, R., and D. Smalling, "A New and Practical Applicaton of Annular Hydraulics," S.P.E. Paper 4518, 1973.

24. Walker, R. E., and D. E. Korry, "Field Method of Evaluating Annular Performance of Drilling Fluids," S.P.E. Paper 4321, 1973.

25. Wallick, G. C., and J. G. Savins, "A Comparison of Differential and Integral Descriptions of the Annular Flow of a Power-Law Fluid," S.P.E. J., September 1969.

26. Wang, Z. Y., and Tang, S. R., "Casson Rheological Model in Drilling Fluid Mechanics," S.P.E. Paper 10564, 1982.

3
THE DRILLING FLUID

3.1 INTRODUCTION

This section presents a limited description of drilling fluids. A complete discussion of drilling fluids (mud types, chemical treatments, solids control, water loss, and so on) is beyond the scope and intent of this manual. A more detailed discussion may be found in Section 4 of the Field Geologist's Training Guide (MS-178).

The drilling fluid performs a variety of functions which includes transporting drilled cuttings to the surface, maintaining borehole stability, balancing formation pressure, cooling the bit, and providing hydraulic energy at the bit to improve penetration rate.

Drilling has been carried out using water, mud, oil, air, mist and foam drilling fluids. Only the behavior of liquid drilling fluids -- water-based muds, emulsions, and oil-based muds -- are considered in this manual.

Each type of drilling fluid consists of three components: the liquid, the active solids, and the inert solids. The active solids react chemically with the liquid to control the fluid properties. In water-based muds, the active solids are usually commercial clays or hydrated clays incorporated from the drilled formation. In emulsions and oil-based muds, the emulsified water acts like an active solid. The inert solids are chemically inactive. The most common inert solids are weighting material and drilled cuttings.

3.2 WATER-BASED MUDS

Water-based muds derive their viscosity from the active solids phase, usually a clay mineral. Sodium montmorillonite, also known as bentonite, is an effective viscosifier in fresh-water muds. It consists of negatively-charged plates, loosely bound by sodium cations. When added to water, the water molecules are absorbed between the plates, causing swelling and developing viscosity. Figure 3-1 illustrates the montmorillonite in water. The platy structure is also effective in controlling the fluid loss and building a good wall cake.

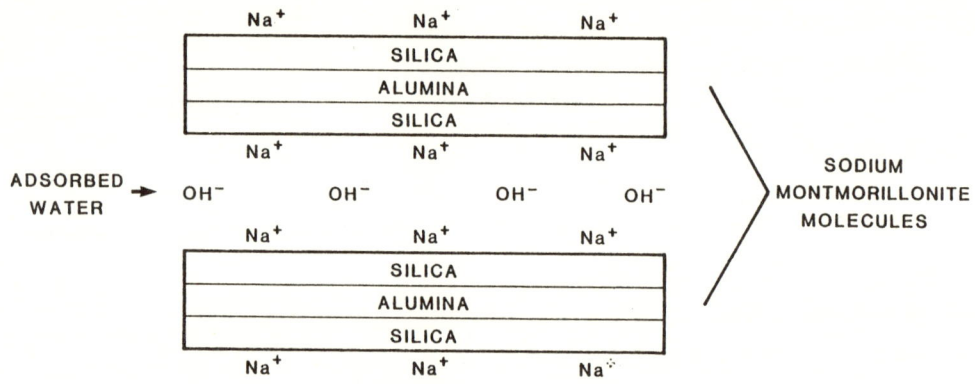

Figure 3-1. Ionic Interaction of Sodium Montmorillonite and Water

Calcium montmorillonite, or sub-bentonite, may also be used for this purpose. Because the calcium ion binds the clay plates more strongly, sub-bentonite swells less and provides less viscosity than sodium montmorillonite. The clay particles are poorly dispersed, and the wall cake is inferior.

Both bentonite and sub-bentonite become ineffective in saline water. The abundant sodium ions bind the clay plates together, causing flocculation. In salt-water muds, the active solid is usually attapulgite, a clay with a needle-like structure. Attapulgite may be used to build viscosity, but it is less effective in controlling fluid loss or building wall cake.

The use of polymers to supplement or replace clays as viscosifiers and fluid loss control agents is becoming increasingly widespread. Most polymers provide a mud with increased shear thinning properties (that is, the viscosity is substantially reduced at high rates of shear). This is a desirable property

3
THE DRILLING FLUID

3.1 INTRODUCTION

This section presents a limited description of drilling fluids. A complete discussion of drilling fluids (mud types, chemical treatments, solids control, water loss, and so on) is beyond the scope and intent of this manual. A more detailed discussion may be found in Section 4 of the Field Geologist's Training Guide (MS-178).

The drilling fluid performs a variety of functions which includes transporting drilled cuttings to the surface, maintaining borehole stability, balancing formation pressure, cooling the bit, and providing hydraulic energy at the bit to improve penetration rate.

Drilling has been carried out using water, mud, oil, air, mist and foam drilling fluids. Only the behavior of liquid drilling fluids -- water-based muds, emulsions, and oil-based muds -- are considered in this manual.

Each type of drilling fluid consists of three components: the liquid, the active solids, and the inert solids. The active solids react chemically with the liquid to control the fluid properties. In water-based muds, the active solids are usually commercial clays or hydrated clays incorporated from the drilled formation. In emulsions and oil-based muds, the emulsified water acts like an active solid. The inert solids are chemically inactive. The most common inert solids are weighting material and drilled cuttings.

3.2 WATER-BASED MUDS

Water-based muds derive their viscosity from the active solids phase, usually a clay mineral. Sodium montmorillonite, also known as bentonite, is an effective viscosifier in fresh-water muds. It consists of negatively-charged plates, loosely bound by sodium cations. When added to water, the water molecules are absorbed between the plates, causing swelling and developing viscosity. Figure 3-1 illustrates the montmorillonite in water. The platy structure is also effective in controlling the fluid loss and building a good wall cake.

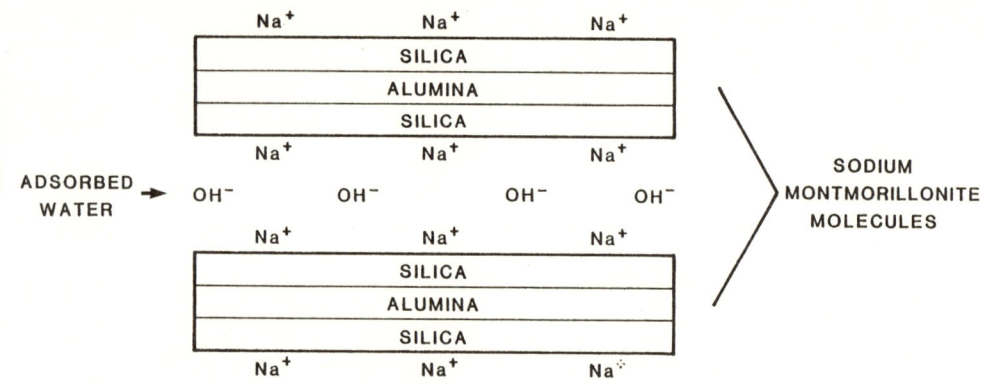

Figure 3-1. Ionic Interaction of Sodium Montmorillonite and Water

Calcium montmorillonite, or sub-bentonite, may also be used for this purpose. Because the calcium ion binds the clay plates more strongly, sub-bentonite swells less and provides less viscosity than sodium montmorillonite. The clay particles are poorly dispersed, and the wall cake is inferior.

Both bentonite and sub-bentonite become ineffective in saline water. The abundant sodium ions bind the clay plates together, causing flocculation. In salt-water muds, the active solid is usually attapulgite, a clay with a needle-like structure. Attapulgite may be used to build viscosity, but it is less effective in controlling fluid loss or building wall cake.

The use of polymers to supplement or replace clays as viscosifiers and fluid loss control agents is becoming increasingly widespread. Most polymers provide a mud with increased shear thinning properties (that is, the viscosity is substantially reduced at high rates of shear). This is a desirable property for a drilling fluid, because viscosity in the annulus can be maintained while the pressure required to circulate is reduced.

3.3   OIL-BASED MUDS

Oil-based muds may be either invert emulsions or true oil muds. In an invert emulsion, the continuous oil phase contains droplets of water. These water droplets act as viscosifiers: viscosity can be increased by increasing the concentration of water in the emulsion. A typical invert emulsion will contain a liquid phase of about 70% oil and 30% water.

True oil muds contain about 5% to 10% water. Frequently, oil muds exhibit lower viscosity than corresponding water-based muds, reducing the pressure required to circulate. Cuttings and weighting material are kept in suspension by the emulsified water.

3.4   MUD VISCOSITY

At very high rates of shear, the interactions between active solid particles become insignificant. Viscosity at high shear rates is controlled and limited by the composition of the liquid phase, the total solids content, and temperature. The plastic viscosity measured by a two-speed viscometer gives an indication of the viscosity at high shear rates, although it has little quantitative meaning unless the fluid follows the Bingham model. Excessive plastic viscosity causes high circulating pressures and can generally be controlled by reducing the solids content of the mud through dilution and mechanical desanders and desilters.

At lower rates of shear, the active solids contribute to the fluid's viscosity. Control of viscosity at low shear rates is particularly important since it affects cuttings transport, suspension of weighting material, and pressure surges applied to the formation through frictional pressures in the annulus. The yield point measured in the field gives an indication of the effectiveness of the active solids. Yield point can be maintained within the desired limits by adjusting the quantity of active solids in the system, and by ensuring that the ionic content of the liquid phase permits the active solids to perform effectively.

3.5 REFERENCES

1. Havenaar, I., "The Pumpability of Clay-Water Drilling Fluids," S.P.E. Pet. Trans. Reprint Series 6 (Drilling).

2. McMordie, W. C., R. B. Bennett and R. G. Bland, "The Effect of Temperature and Pressure on the Viscosity of Oil-Base Muds," J.P.T., July 1975.

3. Moore, P. L., <u>Drilling Practices Manual</u>, Petroleum Publishing Co., 1974.

4
THE MUD CIRCULATING SYSTEM

4.1 INTRODUCTION

A typical drilling rig circulating system is shown in Figure 4-1. The mud pumps draw fluid from the suction pit and pump it through the discharge manifolds, standpipe and kelly hose, and down the drillstring. At the bottom of the hole the fluid passes through the bit nozzles before returning up the annulus and through the flowline and shale shakers to the mud tanks.

4.2 THE PUMPS

Oilfield mud pumps are reciprocating pumps in which fluid is displaced by a piston. Single-acting pumps displace fluid on the forward stroke only, while double-acting pumps displace fluid on both the forward and backward strokes. Most oilfield pumps are either double-acting duplex, with two cylinders, or single-acting triplex, with three cylinders. A rig is typically equipped with two or three pumps so that maintenance can be carried out on one pump without interrupting the drilling operation.

A logging geologist must know how to calculate the volumetric output of a mud pump in order to calculate lag time and circulation time. While the pump output is quoted in the manufacturer's literature and available from the driller or toolpusher, it is not always clear what volumetric efficiency is implied by the quoted output.

Volumetric output is expressed as a volume per stroke. A pump stroke is defined as one revolution of the crankshaft, so that each piston of a pump moves once in each direction during one stroke. As each piston moves forward, it sweeps a volume:

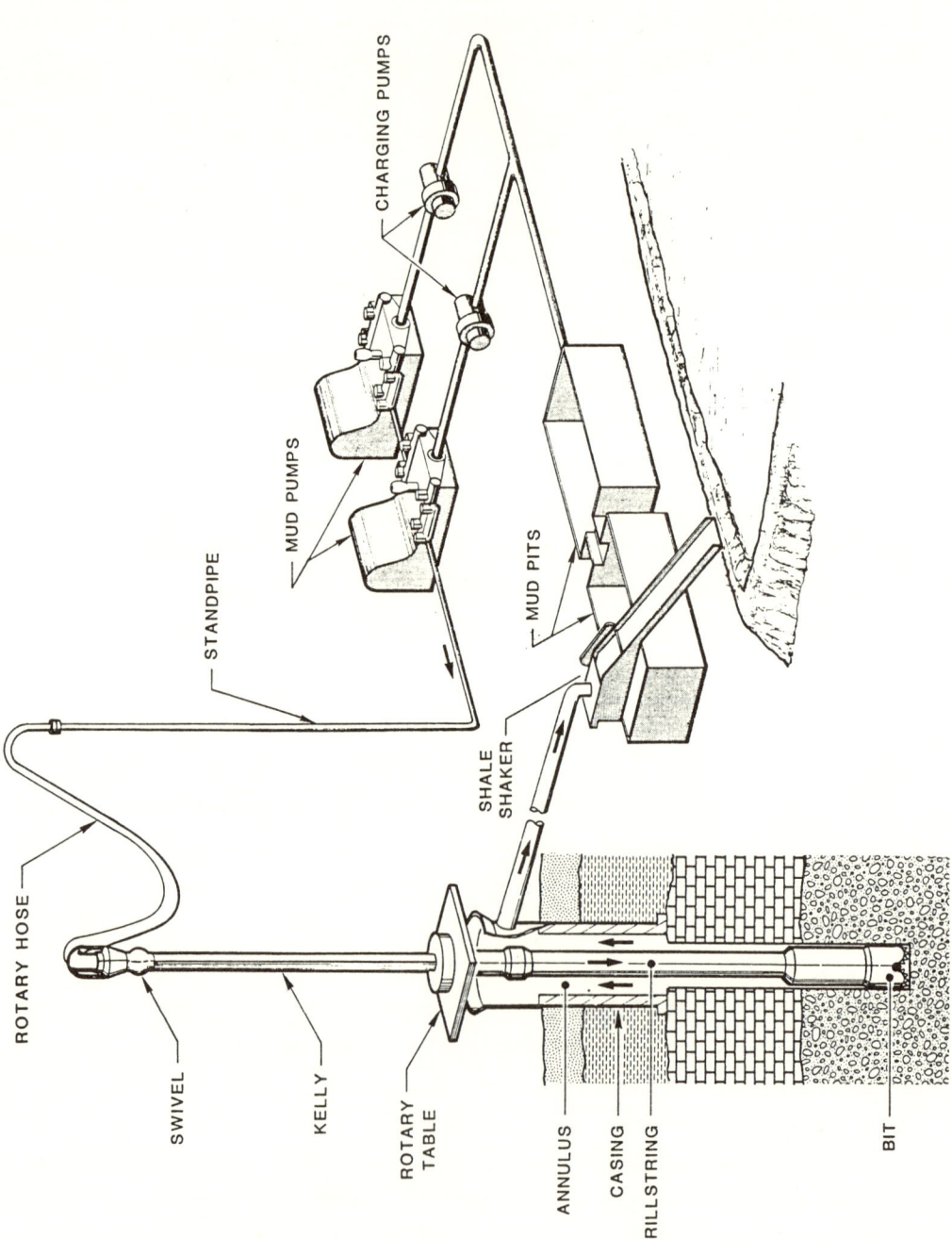

Figure 4-1. The Rig Circulating System

$$L \frac{\pi}{4} D^2$$

As a double-acting piston moves backward, the volume swept is

$$L \frac{\pi}{4} (D^2 - d^2)$$

where

L = length
D = inside diameter of liner
d = outside diameter of rod

Single-acting pumps do not displace fluid on the backward stroke. The volumetric output of a single-acting pump is given by

$$V = LC \xi \frac{\pi}{4} D^2 \qquad (4-1)$$

and the output of a double-acting pump is given by

$$V = LC \xi \frac{\pi}{4} (2D^2 - d^2) \qquad (4-2)$$

where

V = volumetric output
L = cylinder length
C = number of cylinders
ξ = pump efficiency
D = inside diameter of liner
d = outside diameter of rod

Each pump can accommodate a range of liners of differing inside diameters. Smaller liners provide less volumetric output, but they are capable of operating at higher discharge pressures. In designing a hydraulic program it is usual to select the largest liner which will give the required discharge pressure. Unnecessary use of smaller liners requires higher pump speeds, leading to accelerated wear and power losses in the drive train, ultimately causing reduced mechanical efficiency.

Single-acting pumps usually operate at a volumetric efficiency of about 95 percent. If fluid is supplied under positive pressure by supercharging pumps, the efficiency of a single-acting pump may reach 98 percent. The volumetric output of a double-acting pump is usually about 90 percent.

The volumetric efficiency of a pump is not a constant. It can be affected by discharge pressure, pump operating speed, suction line design, fluid aeration, and mechanical wear. High discharge pressures promote leakage and compression of the fluid, thereby reducing volumetric efficiency. High piston speeds, combined with poorly designed suction lines, can cause knocking when fluid cannot enter the pump fast enough to maintain a full cylinder. Knocking causes very rapid mechanical wear. The efficiency of suction lines can be greatly reduced by solids accumulation. Solids accumulation is sometimes indicated by a temperature difference between the top and bottom of the suction line. Suction lines should be as large, short, and straight as possible. Aeration of the drilling fluid causes an increase in compressibility, with a corresponding drop in volumetric efficiency. This can sometimes be seen as a decrease in pump pressure combined with a slight increase in pump rate.

A drop in pump efficiency which is not attributable to the above causes may be the result of leakage past worn valves or seals, indicating that maintenance is required.

The bulk rate of fluid flow is equal to the volumetric output of the pump,

$$Q = V * N_p \qquad (4-3)$$

where

    Q = bulk flowrate
    V = volumetric output
    N_p = stroke rate

4.3 THE DRILLSTRING

From the pumps the drilling fluid flows under pressure through the manifolds, standpipe, kelly hose, drillpipe, and collars. All of these parts of the circulating system may be considered together since they are all essentially circular pipes. Diameters are typically about 0.1 m (4 inches), although inside diameters of drill collars and slim-hole drillpipes are considerably smaller.

The mean or bulk velocity at which the fluid passes through these pipes can be calculated from Equation (4-1):

$$\bar{v} = \frac{Q}{A}$$

$$\bar{v} = \frac{4Q}{\pi D^2} \qquad (4-4)$$

where

$\bar{v}$ = average velocity
Q = bulk flowrate
D = inside diameter of pipe

In typical field situations these velocities are quite high, of the order of 300 m/min (1000 ft/min). At such velocities the fluid is in turbulent flow. Section 5 describes flow in the drillstring in detail, but at this point it will be noted that the pressure required to circulate fluid in turbulent flow varies as approximately the 1.8 power of flowrate; thus, to double the flowrate would increase the pressure drop down the drillstring by approximately $2^{1.8}$, or 3.5 times. Typically, pressure drop down the drillstring represents about 35 percent of total pump pressure.

4.4 THE BIT

The bit nozzles represent a short interval of greatly reduced diameter and increased velocity. Pressure drop through the nozzles is proportional to fluid density and to the square of flowrate. The greatest part of the pressure drop at the nozzles occurs as the fluid exits, which provides mechanical energy to clean the bottom of the hole and the bit. Bit nozzles are designed with sharp exits to maximize this energy. Contrary to some published theories, pressure drop through bit nozzles has very little dependence on friction or on the fluid's viscous properties. In a typical circulating system, the pressure drop across the bit nozzles accounts for about 60 percent of total pump pressure.

4.5 THE ANNULUS

Because the annulus usually has a larger cross-sectional area than the inside of the drillstring, annular velocities are usually lower than internal velocities. Annular velocity must be considered carefully when selecting a flowrate: excessive velocity opposite open hole promotes erosion, while insufficient velocity in the larger annulus near the surface can cause inadequate cuttings transport. Annular velocity is calculated from Equation (2-1):

$$\bar{v} = \frac{Q}{A}$$

$$\bar{v} = \frac{4Q}{\pi(D^2 - d^2)} \qquad (4\text{-}5)$$

The annular flow regime may be laminar, transitional, or turbulent. Laminar flow is usually considered to be desirable when the exposed formation is weak or unstable. In laminar flow, pressure drop is approximately proportional to the flowrate. Pressure loss in the annulus usually represents less than 10 percent of total circulating pressure.

NOTE

Much of this manual will be concerned with calculating the pressures caused by circulating the drilling fluid. In order to simplify the calculations, the minor effects of gravity and atmospheric pressure have been omitted.

One of the major uses of these calculations is for determining the pressures imposed on the open hole by circulating or tripping. It is conventional to express these pressures either as gauge pressures or as an equivalent fluid density, and in either case, the atmospheric pressure may be ignored. The effect of gravity is to add a hydrostatic pressure proportional to the fluid's density and to the vertical extent of the mud column above the point of interest. The hydrostatic pressure is given by Equation (1-1).

Hydrostatic pressure depends only on the vertical height of the mud column, not on the measured length of the column. This height is normally the vertical depth of the hole to the point in question, corrected for the vertical distance from the depth reference datum to the top of the fluid column (the flowline).

While circulating, the pressure imposed at any point in the annulus equals the hydrostatic pressure, plus the sum of the circulating pressure losses from that point to the flowline. This pressure may be converted to an equivalent circulating density by

$$ECD = \frac{P_h + \Sigma P_a}{(D_v - Fl) * g} \qquad (4\text{-}6)$$

or, in oilfield units,

$$ECD = MW + \frac{\Sigma_1 P_a}{(D_v - Fl) * .0519}$$

where

    ECD = Effective Circulating Density
    MW = mud density
    ΣP_a = sum of annular pressure losses
    D_v = vertical depth
    Fl = flowline depth

4.6 REFERENCES

1. Moore, P. L., <u>Drilling Practices Manual</u>, Petroleum Publishing Co., 1974.

2. Randall, B. V., and D. B. Anderson, "Flow of Mud During Drilling Operations," S.P.E. Paper 9444, 1980.

5
THE DRILLSTRING

5.1 INTRODUCTION

The flow of drilling fluid through the kelly hose, kelly, drillpipe and collars may be considered to be flow through a series of circular pipes. Laminar flow in pipes can be analyzed quite readily for most of the fluid models, and a considerable amount of experimental work has been performed on turbulent flow of non-Newtonian fluids in pipes. If the drilling fluid conforms closely to the fluid model in use, the nature of flow in the drillstring can be calculated with confidence.

The engineer needs to know how much pressure will be required to pump fluid through the drillstring at a given rate. It is possible to calculate other parameters such as effective viscosity, Reynolds number and flow regime, but these are not relevant to the drilling operation except as intermediate results in calculating pressure. The remainder of this section will be concerned with the calculation of pressure losses in laminar and turbulent flows.

5.2 LAMINAR FLOW

Laminar flow analysis is largely mathematical and is very similar to the flow analysis in a rotating-sleeve viscometer, presented under paragraph heading 2.19. The general idea is to describe the fluid flow at every point within the conductor. Again, the mathematical analysis is presented separately from the result equations.

5.3 ANALYSIS

The assumptions made for laminar flow in a rotary-sleeve viscometer generally apply to laminar flow in a pipe. It is assumed that

1. The fluid is in laminar flow.
2. The fluid velocity at the pipe wall is zero.
3. The fluid is incompressible.
4. The fluid is completely described by the fluid model.
5. The fluid flow behavior is not time-dependent.
6. The fluid system is in equilibrium.

Figure 5-1 illustrates the model that is assumed for laminar flow analysis. As described under paragraph heading 2.5, laminar flow in a pipe proceeds by the action of concentric fluid shells sliding past one another. Each shell has a constant velocity, and no net force acts upon the system in an equilibrium state. Thus the forces of pressure, which tend to accelerate the fluid shells, are balanced by the viscous drag forces which tend to retard acceleration.

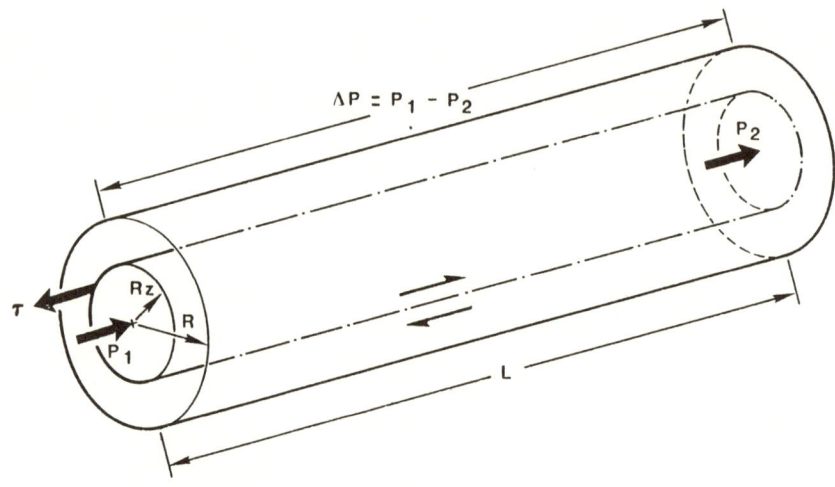

Figure 5-1. Forces in Equilibrium for Laminar Pipe Flow

The pressure differential acts across the top of the shell. Thus

$$F = \Delta P \, \pi R^2 \, z^2 \qquad (5\text{-}1)$$

where

- F = force
- ΔP = pressure differential
- R = radius from the center of the pipe to the inside wall
- z = ratio of radius at some point within pipe to radius of pipe, or $\frac{r}{R}$

The viscous drag due to shear stress acts across the surface area of the shell.

$$F = \tau \, 2\pi R z \qquad (5\text{-}2)$$

where

- F = force
- τ = shear stress
- R = pipe radius
- z = radius ratio

Setting the forces equal to one another yields

$$\Delta P \, \pi R^2 \, z^2 = \tau \, 2\pi R z$$

or

$$\tau = \frac{\Delta P R z}{2L} \qquad (5\text{-}3)$$

Thus the shear stress at any point in the fluid is directly proportional to the radial position at that point. This is true for any fluid model, provided the flow is laminar and in equilibrium. The shear stress is zero at the pipe center, where $z = 0$, and increases to a maximum at the pipe wall. If a fluid has a yield point, there will be a central core of plug flow, the radius of which is $2L\tau_0/\Delta P$, where τ_0 is the yield stress.

Because the shear stress is a function of the radial position, and because shear rate is some function of the shear stress for each model, the shear rate may be expressed as some function of radial distance for each model.

Using the Newtonian fluid model for an example,

$$\tau = \mu * \gamma$$

or,

$$\gamma = \frac{\tau}{\mu} \tag{5-4}$$

From Equations (5-3) and (5-4),

$$\gamma = \frac{\Delta P R z}{\mu 2 L} \tag{5-5}$$

Because shear rate is defined as $\frac{dv}{dr}$, or as $\frac{dv}{Rdz}$ when using the radius ratio z,

$$\frac{dv}{Rdz} = \gamma$$

$$dv = \gamma R dz \tag{5-6}$$

Substituting for shear rate from Equation (5-5),

$$dv = \frac{\Delta P R^2 z}{\mu 2 L} dz \tag{5-7}$$

Therefore,

$$v = \int \frac{\Delta P R^2 z}{\mu 2 L} dz \tag{5-8}$$

Upon moving all constant factors outside the integral and integrating from some point z within the pipe, where z = y (y being a general ratio factor analogous to z), to the pipe wall, where z = 1,

$$v = \frac{\Delta P R^2}{\mu 2 L} \int_y^1 z \, dz \qquad (5\text{-}9)$$

$$v = \frac{\Delta P R^2}{\mu 2 L} \left[\frac{z^2}{2} \right]_y^1$$

$$v = \frac{\Delta P R^2}{\mu 2 L} * .5 \left[z^2 \right]_y^1$$

$$v = \frac{\Delta P R^2}{4 \mu L} (1 - y^2) \qquad (5\text{-}10)$$

The velocity of the fluid at some radial distance (R * y) is expressed by Equation (5-10).

The flowrate of the shell may be found from the flowrate equation

$$Q = \bar{v} * A$$

Therefore

$$dQ = v * 2\pi R y * R dy \qquad (5\text{-}11)$$

The total bulk flowrate is the sum of the flowrates of each individual shell. The total flowrate is obtained by integrating Equation (5-11) from y = 1 at the wall to y = 0 at the center.

$$dQ = v * 2\pi R y * R dy$$

$$Q = 2\pi R^2 \int_0^1 v * y \, dy \qquad (5\text{-}12)$$

Substituting Equation (5-10) for velocity v,

$$Q = 2\pi R^2 \int_0^1 \frac{\Delta P R^2}{4\mu L}(1-y^2)\, y\, dy \qquad (5-13)$$

$$Q = \frac{\Delta P R^2}{4\mu L} * 2\pi R^2 \int_0^1 (y - y^3\, dz)$$

$$Q = \frac{\Delta P \pi R^4}{2\mu L} \left[\frac{y^2}{2} - \frac{y^4}{4}\right]_0^1$$

$$Q = \frac{\Delta P \pi R^4}{2\mu L} (\frac{1}{2} - \frac{1}{4} - 0)$$

$$Q = \frac{\Delta P \pi R^4}{8\mu L} \qquad (5-14)$$

From the flowrate equation,

$$Q = \bar{v} * A$$

$$\bar{v} = \frac{Q}{A}$$

Substituting Equation (5-14) for Q and πR^2 for A,

$$\bar{v} = \frac{\Delta P \pi R^4}{8\mu L * \pi R^2}$$

$$\bar{v} = \frac{\Delta P R^2}{8\mu L} \qquad (5-15)$$

Therefore,

$$\Delta P = \frac{8\mu L \bar{v}}{R^2}$$

5.4 RESULTS

Results of this analysis for a variety of fluid models are as follows:

<u>Newtonian</u>:

$$\Delta P = \frac{8\mu L \bar{v}}{R^2} \qquad (5\text{-}16)$$

<u>Bingham</u>:

$$\left(\frac{2L\tau_o}{\Delta P * R}\right)^4 - 4\left(\frac{2L\tau_o}{\Delta P * R}\right)\left(1 + \frac{3\mu_\infty * \bar{v}}{\tau_o * R}\right) + 3 = 0 \qquad (5\text{-}17)$$

<u>Power Law</u>:

$$\Delta P = \frac{2kL}{R}\left[\frac{\bar{v}(3n+1)}{Rn}\right]^n \qquad (5\text{-}18)$$

<u>Casson</u>:

$$\left(\frac{2L\tau_o}{\Delta P * R}\right)^4 - 28\left(\frac{2L\tau_o}{\Delta P * R}\right)\left(1 - \frac{3\mu_\infty \bar{v}}{\tau_o * R}\right) + 48\left(\frac{2L\tau_o}{\Delta P * R}\right)^{.5} - 21 = 0 \qquad (5\text{-}19)$$

<u>Robertson-Stiff</u>:

$$\left(\frac{2Lk\gamma_o^n}{\Delta P * R}\right)^{3+\frac{1}{n}} - (3n+1)\left(\frac{2Lk\gamma_o^n}{\Delta P * R}\right)^{\frac{1}{n}}\left(1 + \frac{3\bar{v}}{R\gamma_o}\right) + 3n = 0 \qquad (5\text{-}20)$$

5.5 FIELD PROCEDURES

Pressure losses can be obtained directly only for Newtonian or Power Law fluids which exhibit no yield stress. The equation for Bingham fluids can be solved explicitly by the following method:

$$\beta = 1 + \frac{3\mu_\infty \bar{v}}{\tau_0 R} \qquad (5-21)$$

$$z = \left[\beta^2 + \sqrt{\beta^4 - 1}\right]^{1/3} \qquad (5-22)$$

$$y = 2\left(z + \frac{1}{z}\right) \qquad (5-23)$$

$$x = .5\left[\sqrt{y} - \sqrt{\frac{8\beta}{\sqrt{y}} - y}\right] \qquad (5-24)$$

$$\Delta P = \frac{2L\tau_0}{R * x} \qquad (5-25)$$

It is emphasized that the above equations represent exact solutions for laminar pipeflow. Many other solutions may be given in the literature, but in most cases these are approximations -- the validity of which depends on the magnitudes of the quantities involved. The most popular approximations are obtained by neglecting the plug flow region when integrating expressions for fluids with yield stresses, such as Bingham or Robertson-Stiff fluids. For a Bingham fluid, the resulting solution is

$$\Delta P \simeq \frac{8\mu_\infty \bar{v} L}{R^2} + \frac{8\tau_0 L}{3R} \qquad (5-26)$$

or, in oilfield units,

$$\Delta P = \frac{L\mu_\infty \bar{v}}{90{,}000\, D^2} + \frac{L\tau_o}{225\, D}$$

Figure 5-2 shows the magnitude of the error introduced by this popular approximation. The exact solution given in Equations (5-21) through (5-25) is to be preferred if a computer or programmable calculator is available.

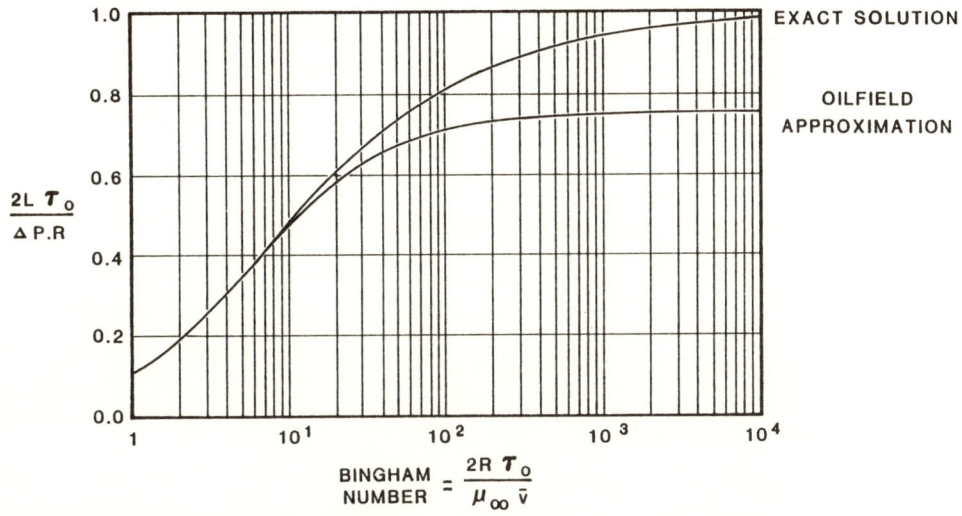

Figure 5-2. Approximation Error for Bingham Fluid

The dimensionless group,

$$\frac{2R\tau_o}{\mu_\infty \bar{v}}$$

plotted along the horizontal axis of Figure 5-2, commonly occurs when considering fluids with yield stress. In studies of Bingham fluids, it has been called the "Bingham number" or "plasticity." The term "plasticity" is more useful since it can also be applied to Casson fluids, or, in the form $(2R\gamma_o)/\bar{v}$, to Robertson-Stiff fluids.

5.6 THE REYNOLDS NUMBER

The Reynolds number is a dimensionless, empirically-deduced parameter. The significance of the Reynolds number is in its use as a correlation parameter: different fluids with different properties exhibit similar flow characteristics at the same Reynolds number.

The major use of the Reynolds number is the determination of flow regime. Generally, the flow regime of a liquid changes from laminar to turbulent at a fairly well-defined Reynolds number value.

The Reynolds number was the result of the experiments in 1883 of Osborne Reynolds. Using water as the subject fluid, Reynolds related the various factors affecting flow (fluid density, fluid viscosity, the average velocity, and pipe diameter) and defined the Reynolds number as

$$Re = \frac{\rho \overline{v} D}{\mu} \qquad (5-27)$$

or, in oilfield units

$$Re = \frac{15.47 * MW * \overline{v} * D}{\mu}$$

Reynolds observed that water changed from laminar flow to turbulent flow at a Reynolds number of approximately 2000.

It is apparent that Equation (5-27) is not valid for a non-Newtonian fluid because non-Newtonian fluids do not have an absolute viscosity; the viscosity varies with shear rate. For non-Newtonian fluids, a viscosity called "mean viscosity" must be used in Equation (5-27). Mean viscosity is defined as the viscosity of a Newtonian fluid which, in laminar flow, would develop the same pressure loss as the subject fluid. Thus from Equation (5-16),

$$\Delta P = \frac{8 \mu L \overline{v}}{R^2}$$

Therefore,

$$\bar{\mu} = \frac{\Delta P_L R^2}{8L\bar{v}} \qquad (5-28)$$

or

$$\bar{\mu} = \frac{\Delta P_L D^2}{32L\bar{v}}$$

where

$\bar{\mu}$ = mean viscosity
ΔP_L = laminar pressure loss

The laminar pressure losses for non-Newtonian fluids are calculated from Equations (5-17) to (5-20).

Substituting mean viscosity from Equation (5-28) for viscosity in (5-27) gives the Reynolds number for a non-Newtonian fluid, also called the equivalent Reynolds number.

$$Re_e = \frac{\rho \bar{v} D}{\bar{\mu}}$$

$$Re_e = \frac{\rho \bar{v} D * 32L\bar{v}}{\Delta P_L D^2}$$

$$Re_e = \frac{32 \rho \bar{v}^2 L}{\Delta P_L D} \qquad (5-29)$$

where

Re_e = equivalent Reynolds number

Generally, for non-Newtonian fluids, the transition from laminar flow to turbulent flow in a pipe occurs also at a Reynolds number of approximately 2000. More-detailed discussions of the transition region are given under paragraph headings 2.7 and 5.9.

5.7 CRITICAL VELOCITY

Another value used to determine flow regime is the critical velocity. The critical velocity is calculated by solving the Reynolds-number equation for velocity at the Reynolds-number value at which turbulence begins, called the critical Reynolds number. Thus, from Equation (5-27), the critical velocity of a Newtonian fluid is

$$v_c = \frac{Re_c \mu}{\rho D} \qquad (5\text{-}30)$$

where

v_c = critical velocity
Re_c = critical Reynolds number

The critical velocity for a non-Newtonian fluid is obtained from Equation (5-29).

$$v_c = \left(\frac{Re_c \Delta P_L * D}{32 \rho L} \right)^{.5} \qquad (5\text{-}31)$$

For the more complex fluid models, critical velocity cannot be obtained directly because ΔP_L is a function of v_c. However, it can be found by iterative procedures.

5.8 TURBULENT FLOW

Unlike laminar flow, the analysis of fluid-flow pressure losses in a pipe in the turbulent regime is largely empirical. The random shearing and intermixing motion of the fluid particles makes orderly mathematical analysis nearly impossible. Pressure losses in turbulent flow are calculated from the Fanning equation, defined for any fluid model by

$$\Delta P = \frac{2f \, L \rho \bar{v}^2}{D} \qquad (5\text{-}32)$$

where

ΔP = pressure loss
f = friction factor
L = length of section
ρ = fluid density
$\bar{v}$ = average velocity
D = pipe diameter

The Fanning equation is empirically derived, and, like the Reynolds number, it attempts to quantify the basic factors affecting flow.

The parameter f, called the Fanning friction factor, is a function of the Reynolds number and of the surface conditions of the pipe wall. The surface condition of a pipe wall is given by the relative roughness parameter $\frac{\varepsilon}{D}$. ε, or the absolute roughness, is the average depth of the pipe wall irregularities. D is the inside pipe diameter. The parameters are illustrated in Figure 5-3. Notice that the smoother the pipe, the lower the value $\frac{\varepsilon}{D}$. As one would expect, lower values of relative roughness are reflected by lower friction factor f which, in turn, results in lower pressure losses.

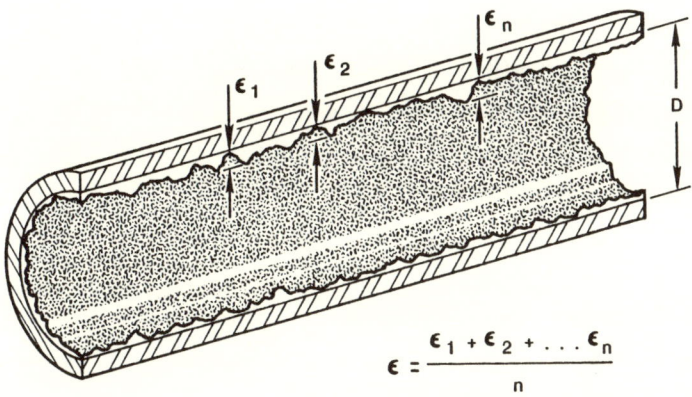

$$\varepsilon = \frac{\varepsilon_1 + \varepsilon_2 + \ldots \varepsilon_n}{n}$$

Figure 5-3. Relative Roughness of a Pipe

Values for the friction factor are derived from both theoretical analysis and empirical data. According to the Fanning equation,

$$f = \frac{\Delta PD}{2L\rho \overline{v}^2} \qquad (5\text{-}33)$$

In laminar flow, for a Newtonian fluid the pressure drop ΔP is defined by Equation (5-16). Substituting yields

$$f = \frac{32\mu L \overline{v}}{D^2} * \frac{D}{2L\rho \overline{v}^2}$$

$$f = \frac{16\mu}{\rho \overline{v} D}$$

$$f = \frac{16}{Re} \qquad (5\text{-}34)$$

Equation (5-34) defines the friction factor in terms of Reynolds number for a Newtonian fluid in laminar flow.

In turbulent flow, the pressure drop (ΔP) in Equation (5-33) is not known. Pressure losses for a fluid with known flow properties are experimentally measured, and the friction factor may then be calculated. Because the flow properties are known, the Reynolds number is known. Thus friction factors are plotted against Reynolds numbers for flow in a pipe of a known relative roughness.

Figure 5-4 illustrates the relationship between friction factor and Reynolds number for a Newtonian fluid. The laminar flow regime is represented by the straight line, defined by Equation (5-34). In turbulent flow, a series of curves are generated by evaluating the relationship at various values of relative roughness of the pipe. Relative roughness values are generally not directly measured at the wellsite. If a relative roughness value is not specified by the drilling engineer, then a value of zero is assumed.

For a non-Newtonian fluid, the curves of Figure 5-4, though derived for a Newtonian fluid, will yield accurate values for the friction factor when an equivalent Reynolds number is used. This is due to the fact that the friction factor is not a very strong function of viscosity.

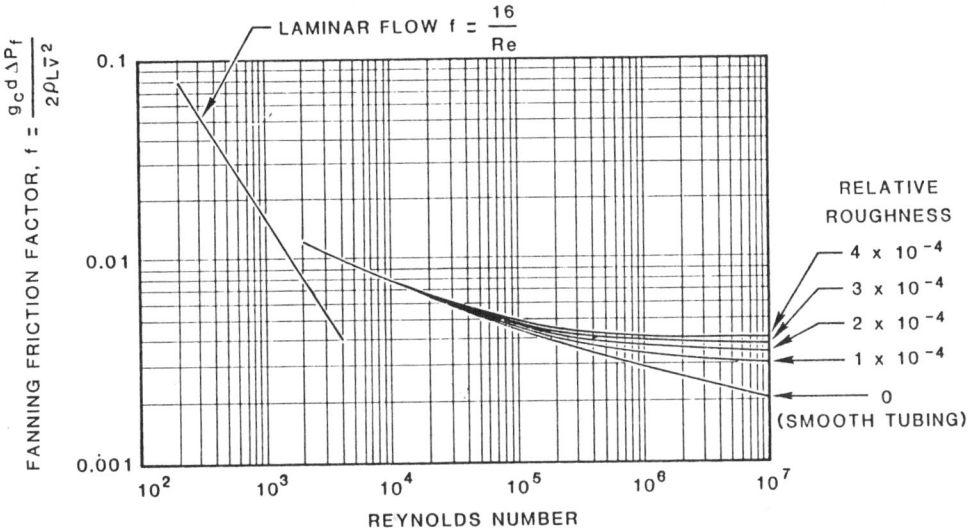

Figure 5-4. Friction Factor Curves for Newtonian Fluid, from the Colebrook Equation

Several mathematical relationships between the Reynolds number and the friction factor have been developed. The curves of Figure 5-4 were derived from the Colebrook equation for Newtonian fluids, given by

$$\frac{1}{\sqrt{f}} = -4 \log \left[\frac{\varepsilon}{3.72\ D} + \frac{1.255}{Re\sqrt{f}} \right] \qquad (5-35)$$

The Blasius relationship is of the form

$$f = y * Re^{-z} \qquad (5-36)$$

which is a power function, giving a straight line on a log-log plot. For Newtonian fluids, values of y range from .046 to .079 and values of z range from .20 to .25. For Power Law fluids, y and z show some dependence on the flow behavior index n. The experimental results of Dodge and Metzner, analyzed by Schuh, yield expressions for y and z.

$$y = \frac{(\log n + 3.93)}{50} \tag{5-37}$$

$$z = \frac{(1.75 - \log n)}{7} \tag{5-38}$$

The index n is calculated from Equation (2-29).

The von Karman relationship is more complex, of the form

$$\frac{1}{\sqrt{f}} = y \log \mathrm{Re} \sqrt{f} + z \tag{5-39}$$

The use of this relationship with non-Newtonian fluids was investigated by Dodge and Metzner who proposed the following modification for Power Law fluids:

$$\frac{1}{\sqrt{f}} = \frac{4}{n * 75} * \log \left[\mathrm{Re} * f^{1-(\frac{n}{2})} \right] - \frac{.4}{n^{1.2}} \tag{5-40}$$

Figures 5-5 and 5-6 show the Blasius and von Karman correlations for various Power Law fluids. A line is also drawn to represent the laminar flow regime.

The mathematical expressions given above are widely used for computerized analysis of turbulent flow. Mathematical expressions for other non-Newtonian fluids besides the Power Law fluid are sparse because the preparation of fluids with the necessary properties is difficult.

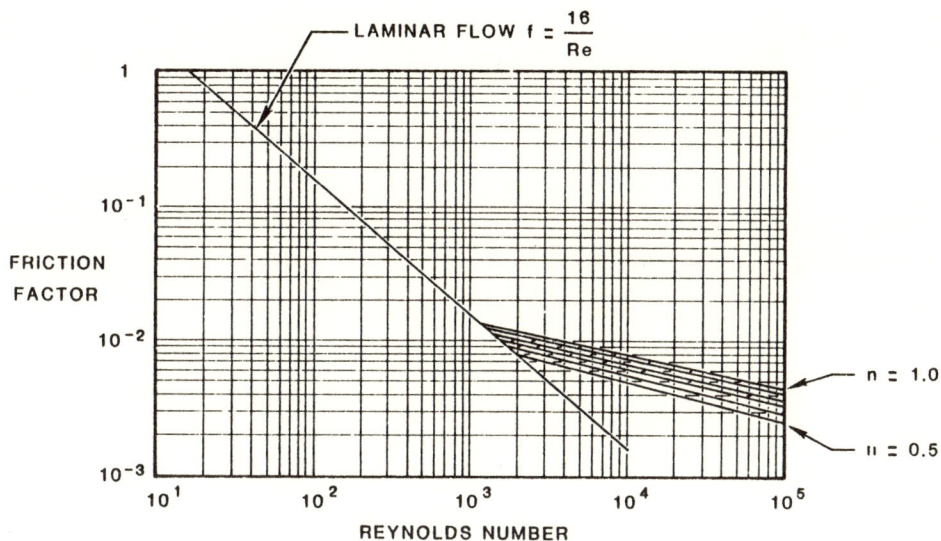

Figure 5-5. Blasius Correlation for Various Power Law Fluids: $\varepsilon/P = 0$

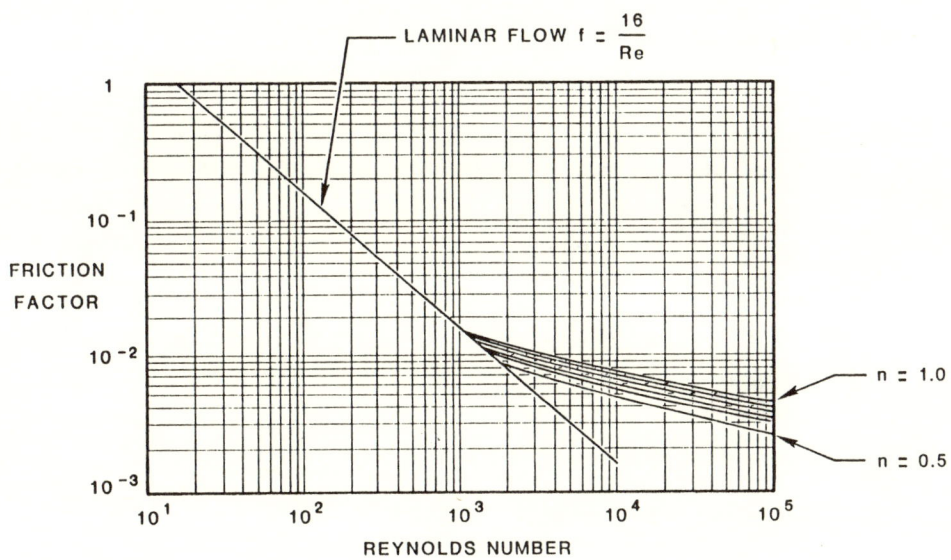

Figure 5-5. Von Karman Correlation for Various Power Law Fuids: $\varepsilon/D = 0$

5.9 TRANSITIONAL FLOW

Calculation of pressure losses in laminar or turbulent flow has now been covered. It is also necessary to determine at what point the flow regime changes from laminar to turbulent. As shown in Figure 5-4, this occurs at a Reynolds number somewhat higher than the intersection of the laminar and turbulent friction factors. That is, laminar flow persists into the region where laminar pressure drop is less than turbulent pressure drop.

The degree to which laminar flow can persist is highly dependent on physical conditions. Turbulence is promoted by rough or wavy walls, increases in pipe diameter, flow variations and vibration -- all conditions which exist in the drillstring. Most experiments on the laminar-to-turbulent transition have been performed under laboratory conditions, and it is reasonable to assume that turbulence is induced somewhat earlier in the drillstring than is suggested by these experiments.

The onset of turbulence is not immediate. There is a transitional zone in which flow is neither laminar nor fully turbulent, and in which observed pressure drops are intermediate between those for laminar and turbulent flow. Based on the experiments of Dodge and Metzner, Schuh has determined the lower limit of the transitional flow regime to occur at a Reynolds number of (3470 - 1370 n) for Power Law fluids. The transitional flow regime extended over a Reynolds number range of 800, its upper limit being (4270 - 1370 n).

For non-Power Law fluids, it is common to assume that turbulence begins at a Reynolds number of 2000. Metzner and Reed suggest that the criterion should be a laminar friction factor of 0.008, which corresponds to Re = 2000 for all fluids, if Re is defined according to Equation (5-29). (Metzner and Reed had defined Re differently, however.)

An alternative criterion has been proposed by Ryan and Johnson. They suggest that the laminar-turbulent boundary depends on the value of a stability parameter Z. This is similar to the Reynolds number, but it is evaluated at one point in the flow channel using radius, local velocity and apparent viscosity at that point. Z is zero at the pipe center because D = 0, and Z is zero at the pipe wall because $\bar{v}$ = 0; a maximum value of Z is found at some intermediate point between the center and wall of the pipe. It is proposed that if maximum Z exceeds 808, the fluid will be in turbulent flow. This criterion has sometimes been applied to oilfield problems, although the Reynolds number correlation is more commonly used. In a Newtonian fluid, the criterion Z_{max} = 808 is equivalent to Re = 2100.

For Bingham fluids in turbulent flow, some data is provided by Hanks and Pratt who recommend the following method for locating the critical Reynolds number:

1. Calculate the Hedstrom number, which is a dimensionless number that characterizes a Bingham fluid.

$$He = \frac{\rho D^2 \tau_0}{\mu_\infty^2} \qquad (5-41)$$

   where

   He = Hedstrom number
   ρ = fluid density
   D = pipe diameter
   τ_0 = yield stress
   μ_∞ = plastic viscosity

2. Solve a cubic equation to determine dimensionless reciprocal pressure drop, x, at the critical point, x_c:

$$16{,}800\, x_c = He(1 - x_c)^3 \qquad (5-42)$$

3. Determine critical Reynolds number from

$$Re_c = \frac{He}{8 x_c} * \left[1 - \frac{4 x_c}{3} + \frac{x_c^4}{3} \right]^2 \qquad (5-43)$$

This gives a critical Reynolds number as defined elsewhere in this manual. The expression given by Hanks and Pratt differs slightly because they define Reynolds number differently. The above calculations are consistent with their work and with the rest of this manual. As the Hedstrom number approaches zero, the limiting value of Re_c is 2100 for a Newtonian fluid, as illustrated in Figure 5-7.

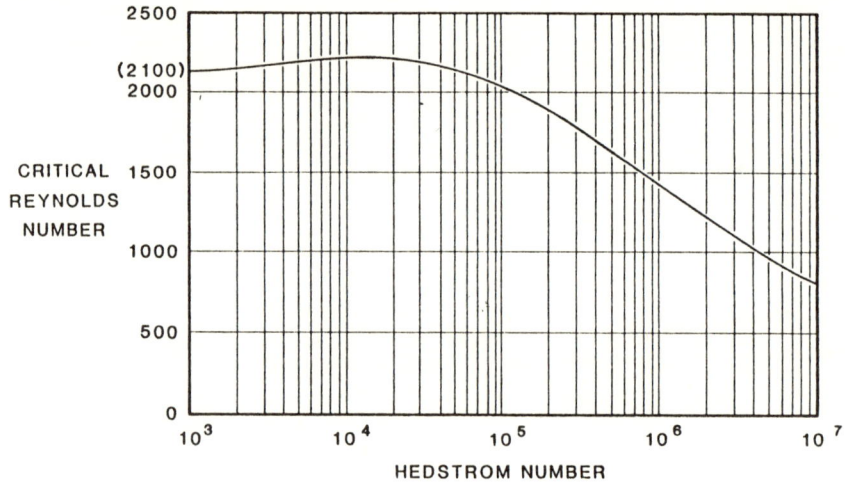

Figure 5-7. Critical Reynolds Number for Pipe Flow, by Method of Hanks and Pratt

5.10 TOOL JOINTS

Most drillpipe in common usage has a reduced inside diameter, or internal upset, at each tool joint. The pressure drop caused by such a restriction can be calculated by standard engineering methods. The pressure loss as the fluid enters the restriction is

$$\Delta P_i = \frac{C_i \rho}{2} \left(v_2^2 - v_1^2 \right) \tag{5-44}$$

where

ΔP_i = pressure loss at restriction entrance
C_i = inlet coefficient
ρ = fluid density
v_2 = velocity within restriction
v_1 = velocity outside restriction

The pressure loss at the outlet is

$$\Delta P_o = \frac{C_o \rho}{2} \left(v_2^2 - v_1^2 \right) \tag{5-45}$$

where

ΔP_o = pressure loss at restriction outlet
C_o = outlet coefficient
v_2 = velocity within restriction
v_1 = velocity outside restriction

Values for C_i and C_o have been empirically determined. Values for C_i range from 0.5 for a sharp-edged orifice to 0.05 for a smooth bell mouth, while C_o is approximately double C_i. Assuming that $(C_o + C_i) \simeq 1$, the total pressure loss at one tool joint is

$$\Delta P_j \simeq \frac{\rho \left(v_2^2 - v_1^2 \right)}{2} \tag{5-46}$$

This pressure should be multiplied by the number of tool joints in the drillstring to find the total pressure drop caused by the tool joint upsets.

5.11 REFERENCES

1. Dodge, D. W., and A. B. Metzner, "Turbulent Flow of non-Newtonian Systems," A.I.Ch.E. J. $\underline{5}$ (2), June 1959.

2. Hanks, R. W., and D. R. Pratt, "On the Flow of Bingham Plastic Slurries in Pipes and Between Parallel Plates," S.P.E. J., December 1967.

3. Havenaar, I., "The Pumpability of Clay-Water Drilling Fluids," S.P.E. Pet. Trans. Reprint Series 6 (Drilling).

4. Howell, J. N., "Improved Method Simplifies Friction-Pressure-Loss Calculations," O&G J., April 25, 1966.

5. Krieger, I. M., and J. S. Dodge, "The Laminar-Turbulent Transition in Suspension of Rigid Spheres," S.P.E. J., September 1967.

6. Lord, D. L., B. W. Hulsey and L. L. Melton, "General Turbulent Pipe Flow Scale-Up Correlation for Rheologically Complex Fluids," S.P.E. J., September 1967; and Discussion, S.P.E. J., March 1969.

7. Melrose, J. D., J. G. Savins, W. R. Foster and E. R. Parish, "A Practical Utilization of the Theory of Bingham Plastic Flow in Stationary Pipes and Annuli," Pet. Trans. A.I.M.E. $\underline{213}$, 1958.

8. Melton, L. L., and W. T. Malone, "Fluid Mechanics Research and Engineering Application in non-Newtonian Fluid Systems," S.P.E. J., March 1964.

9. Metzner, A. B., and J. C. Reed, "Flow of non-Newtonian Fluids -- Correlation of the Laminar, Transition, and Turbulent-Flow Regions," A.I.Ch.E. J., $\underline{1}$ (4), December 1955.

10. Moore, P. L., <u>Drilling Practices Manual</u>, Petroleum Publishing Co., 1974.

11. Randall, B. V., and D. B. Anderson, "Flow of Mud During Drilling Operations," S.P.E. Paper 9444, 1980.

12. Ryan, N. W., and M. M. Johnson, "Transition from Laminar to Turbulent Flow in Pipes," A.I.Ch.E. J., $\underline{5}$ (4), December 1959.

13. Savins, J. G., "Generalized Newtonian (Pseudoplastic) Flow in Stationary Pipes and Annuli," Pet. Trans. A.I.M.E. $\underline{213}$, 1958.

14. Savins, J. G., G. C. Wallick and W. R. Foster, "The Differentiation Method in Rheology - I, Poiseuille Type Flow," S.P.E. J., September 1962.

15. Savins, J. G., G. C. Wallick and R. W. Foster, "The Differentiation Method in Rheology - II, Characteristic Derivatives of Ideal Models in Poiseuille Flow," S.P.E. J., December 1962.

16. Schuh, F. J., "Computer Makes Surge Pressure Calculations Useful," O&G J., August 3, 1964.

17. Wang, Z. Y., and S. R. Tang, "Casson Rheological Model in Drilling Fluid Mechanics," S.P.E. Paper 10564, 1982.

18. Zamora, M., and D. L. Lord, "Practical Analysis of Drilling Mud Flow in Pipes and Annuli," S.P.E. Paper 4976, 1974.

6
MOTOR, TURBINE, & BIT

6.1 INTRODUCTION

An understanding of fluid flow through positive displacement motors, turbines, bits, and bit nozzles is of great importance in determining the performance and efficiency of the drilling assembly. It must be made clear at the outset that there is a vast difference in operating characteristics between a positive displacement motor, also known as a Moineau motor, and a turbine or turbodrill. Well-known examples of Moineau motors are the Christensen Navi-Drill, the Smith Dyna-Drill, and the Eastman Whipstock PDM motor. The Neyrfor turbodrills and the Eastman Whipstock turbine are examples of turbines.

6.2 THE MOTOR

The Moineau motor, illustrated in Figure 6-1, operates by forcing the drilling fluid between a helical motor and a sealing stator. A known and constant amount of rotation is required to pass a fixed volume of fluid through the system, so the motor rpm is proportional to flowrate. Because any resistance to turning causes an increase in pressure, the pressure drop across the motor is proportional to rotary torque. A bypass valve prevents excessive pressures from causing damage, and the pressure at which this valve opens is known as the stall pressure. The bypass valve is also opened while tripping in and out of the hole, allowing the drillstring to fill and drain, and reducing swab and surge pressures.

The power absorbed by a motor is the pressure drop across the motor, multiplied by the flowrate. The rotary power developed is the torque multiplied by the rotary speed. The ratio of these quantities gives the mechanical or hydraulic efficiency of the motor. Moineau motors exhibit almost constant effi-

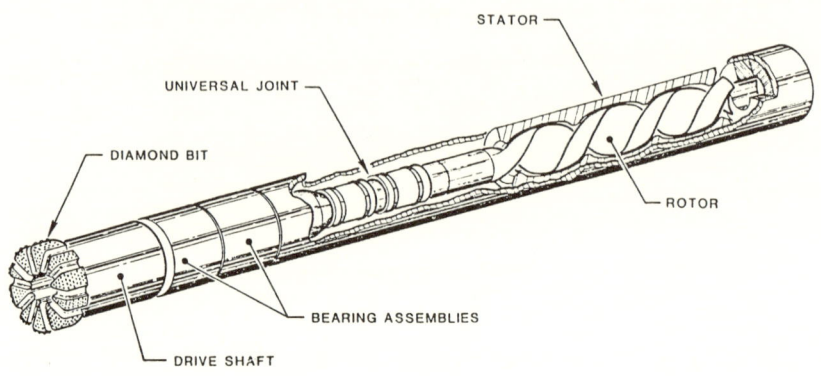

Figure 6-1. Moineau Motor

ciency of about 80 percent over their usual operating range of pressures and flowrates. Thus,

$$\text{hydraulic efficiency} = 80\% = \frac{\text{power absorbed}}{\text{power developed}}$$

$$.80 = \frac{\Delta P * Q}{\text{torque} * N}$$

$$\text{torque} = \frac{\Delta P}{.80} * \frac{Q}{N} \qquad (6\text{-}1)$$

where

ΔP = pressure drop
Q = bulk flowrate
N = rotary speed (rpm)

The quantity $\frac{Q}{N}$ is specified for the motor in gal/min and is constant. Because the torque of a downhole motor can not be measured, Equation (6-1) is not used to calculate the pressure drop across the motor. Pressure drop is estimated by subtracting all other pressure losses in the circulating system from the total pressure observed at the standpipe. The torque may then be estimated by using Equation (6-1).

6.3 THE TURBINE

Unlike a Moineau motor, a turbine (illustrated in Figure 6-2) operates at approximately constant pressure drop for a given fluid density and flowrate. The pressure drop is proportional to the fluid density and to the square of the flowrate, and the manufacturer's specifications usually quote a nominal pressure drop at one specified density and flowrate. Knowing actual density and flowrate, it is therefore possible to calculate the pressure drop across a turbine, provided that it is operating within normal limits. It is not usually possible to calculate torque, rpm, or efficiency with any degree of accuracy. The efficiency of a turbine depends on the ratio of rpm to fluid flowrate, and is extremely variable as rpm changes according to the applied torque.

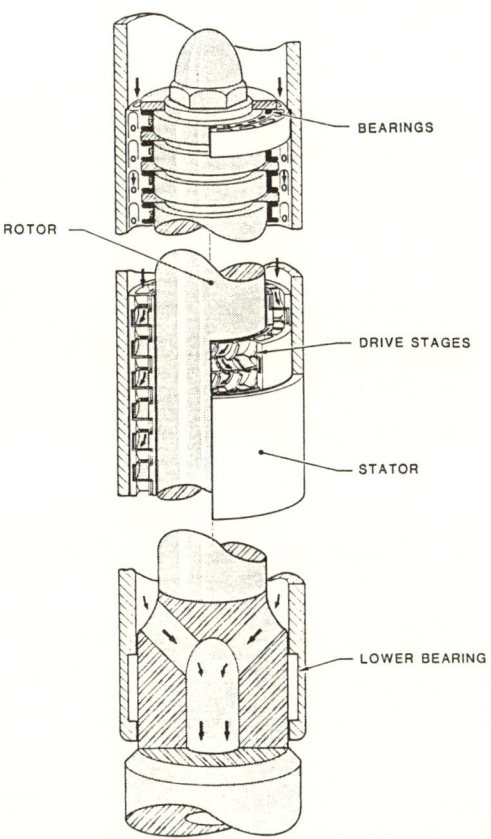

Figure 6-2. Turbine Motor

6.4 THE BIT NOZZLES

Conventional bits incorporate a number of nozzles (usually three) through which the drilling fluid is forced at high velocity. The jet velocity can be calculated by

$$v_j = \frac{Q}{\Sigma A_n} \qquad (6\text{-}2)$$

where

v_j = jet velocity
Q = bulk flowrate
A_n = total area of nozzles

Even if the jet nozzles are of differing sizes, the velocity is the same through each. Fluid velocity through an orifice is determined only by the total pressure drop across the orifice, the fluid density, and the nozzle coefficient. The velocity is independent of the individual diameters of the nozzles.

The standard formula for pressure loss through an orifice is

$$\Delta P_{or} = \left(\frac{1}{C_v^2} - 1\right) \frac{\rho v_j^2}{2} \qquad (6\text{-}3)$$

where

ΔP_{or} = pressure loss through an orifice
C_v = velocity coefficient
ρ = fluid density
v_j = jet velocity

and the pressure loss caused by turbulence at the orifice outlet is

$$\Delta P_t = C_o * \frac{\rho v_j^2}{2} \qquad (6\text{-}4)$$

where

ΔP_t = pressure loss due to turbulence
C_o = outlet coefficient
ρ = fluid density
v_j = jet velocity

The coefficients C_o and C_v are empirically deduced and depend upon the shape of the orifice. The value of C_o for a fixed-nozzle square-edged outlet is 1. Thus, summing Equations (6-3) and (6-4),

$$\Delta P_b = \frac{1}{C_v^2} * \frac{\rho v_j^2}{2} \qquad (6-5)$$

where
ΔP_b = total pressure loss across the bit

For bit nozzles it is assumed $C_v = 0.95$.

Notice that, by substituting the above values, Equations (6-4) and (6-5) indicate that approximately 90 percent of the total pressure loss at the bit is due to the turbulence at the outlet, and only 10 percent is due to pressure loss through the orifice. The bit nozzles are designed to maximize the portion of pressure loss that may be used to perform work on the formation.

Calculation of pressure drop across the bit has a number of important practical applications.

- It enables the engineer to select nozzle sizes to optimize the hydraulic power or impact force at the bit. This improves chip removal and bottom-hole cleaning.

- Together with calculation of drillstring and annular losses, it gives a means of comparing observed standpipe pressure with the total calculated pressure drop; thus the pump's volumetric efficiency can be estimated. A drop in volumetric efficiency usually gives early warning of pump failure.

Pressure drop across unconventional bits such as polycrystalline diamond bits Stratapax-type) may be calculated by Equations 6-2 and 6-5. There will probably be more nozzles than in a conventional bit, but the same method applies.

Unfortunately, it is not possible to calculate the pressure drop developed across the more familiar diamond drilling or core bit. Pressure drop is a function not only of the bit design and circulating parameters, but also of weight-on-bit and formation characteristics. The bit pressure drop can be estimated by subtracting all other calculated pressures from the standpipe pressure; however, even this method does not apply if the bit is run below a positive displacement motor.

The following table summarizes the calculations which can be made for various combinations of bit type and rotary drive:

Bit Type	Rotation	Bit Pressure Drop	Torque	RPM	Pump Efficiency
Conventional	Rotary table	Calculated	Measured	Measured	Estimated
Conventional	Moineau	Calculated	Estimated	Calculated	Estimated
Conventional	Turbine	Calculated	Unknown	Unknown	Estimated
Diamond	Rotary table	Estimated	Measured	Measured	Unknown
Diamond	Moineau	Unknown	Unknown	Calculated	Unknown
Diamond	Turbine	Estimated	Unknown	Unknown	Unknown

6.6 REFERENCES

1. Allen, J. H., and W. Baker, "Improved Drilling Performance with Enhan- Crossflow Rock Bits," J.P.T., October 1980.

2. Herbert, P., "Turbodrilling in the Hot-Hole Environment," J.P.T., Octo- 1982.

3. McLean, R. H., "Crossflow and Impact Under Jet Bits," J.P.T., November 1964

4. Moore, P. L., *Drilling Practices Manual*, Petroleum Publishing Co., 1974.

5. Myers, H. M., and J. E. Funk, "Fluid Dynamics in a Diamond Drill Bit" S.P.E. J., December 1967.

6. Sutko, A. A., and G. M. Myers, "The Effect of Nozzle Size, Number and Extension on the Pressure Distribution Under a Tricone Bit," J.P.T., November 1971.

7. Tirapolsky, W., "Proper Pipe Rotation Enhances Downhole Motor Performance," World Oil, June 1980.

7
THE ANNULUS

7.1 INTRODUCTION

Accurate calculation of annular flow characteristics is of great importance. The determination of effective circulating density, cuttings transport, flow regime, and erosive potential all depend upon proper analysis of annular flow.

Annular flow analysis is very similar to pipe flow analysis. The principles of analysis are identical, and ultimately both are primarily concerned with pressure losses. The determination of annular pressure losses requires the determination of a Reynolds number and a mean viscosity for non-Newtonian fluids. Pressure loss depends upon regime type, so each regime will be analyzed separately.

The major differences in the analyses are due to the differing geometries of the conduit under consideration. The necessary integrations over the annular pipe region are not readily solvable for many fluid models; therefore, exact solutions for pressure loss are difficult to obtain. However, good approximate solutions are achieved by considering annular flow to be nearly identical to flow between parallel plates. Exact solutions and approximate solutions are both discussed.

7.2 LAMINAR FLOW

7.3 EXACT SOLUTIONS

Figures 7-1 and 7-2 show a profile for a Newtonian fluid and a fluid with a yield stress flowing in the annulus. From the two figures one can see that the relationship between fluid velocity and radial position in an annulus is more complex than for a pipe.

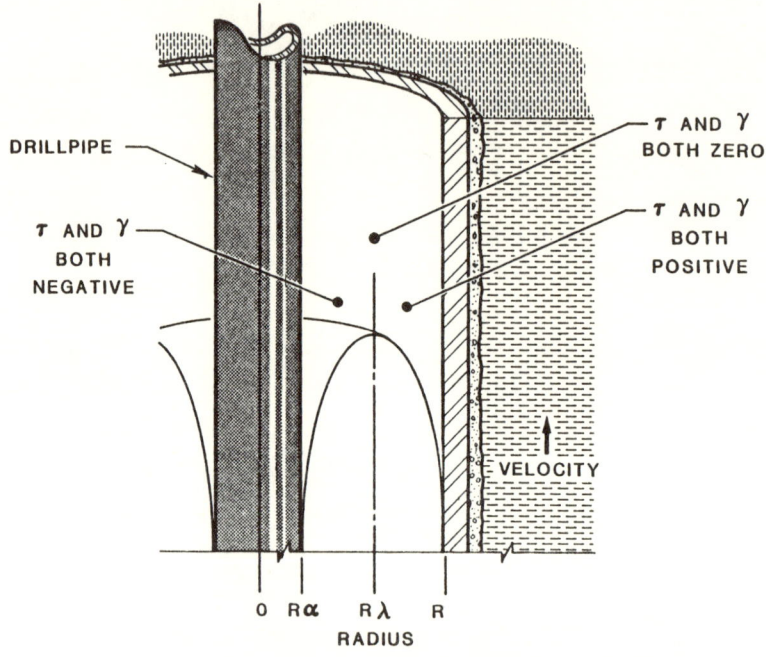

Figure 7-1. Velocity Profile of Laminar Flow in an Annulus, For a Newtonian Fluid

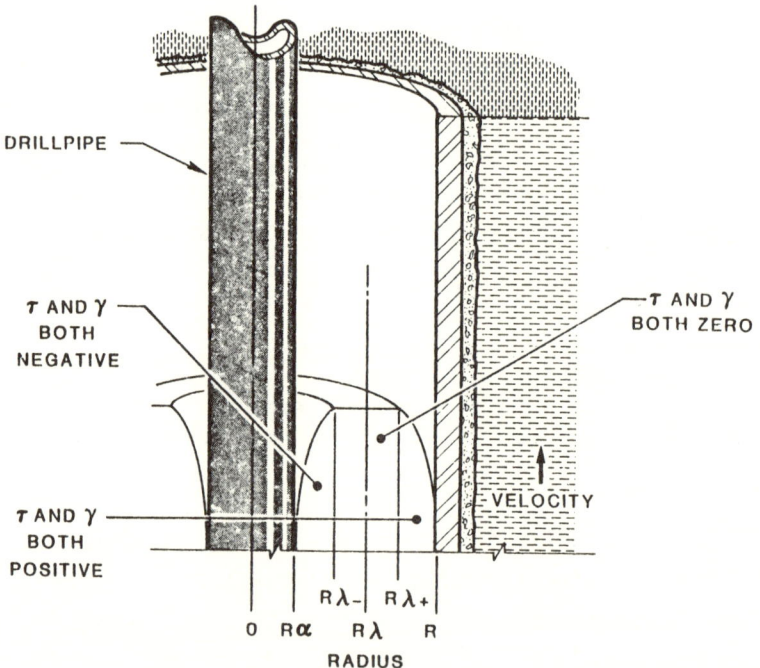

Figure 7-2. Velocity Profile of Laminar Flow in an Annulus, For a Fluid with a Yield Stress

Again, y is a ratio of some radius within the annulus to the radius of the casing or open hole, α is the ratio of the outside radius of the drillpipe to the radius of the casing or the open hole, and λ is the ratio of the radius of maximum fluid velocity to the radius of the casing or open hole. For the Newtonian fluid, in the region bounded by $\alpha \leq y \leq \lambda$, the shear rate and shear stress are negative because velocity difference is negative for a positive change in radius. Conversely, in the region $\lambda \leq y \leq 1$, the shear rate and shear stress are positive because velocity difference is positive for a positive change in radius.

For a fluid with a yield stress, there will be a plug flow region with zero shear rate and zero shear stress in the zone given by $\lambda_- \leq y \leq \lambda_+$.

Any fluid model can be described in general terms as

$$\gamma = f(\tau) \tag{7-1}$$

Figure 7-3 illustrates the model that is assumed for the analysis of laminar flow in an annulus. The same basic assumption for pipe flow applies to annular flow. The fluid is in equilibrium, and analysis proceeds in the same manner. The force moving the fluid shells in the direction of flow is in equilibrium with the force resisting the flow. Thus

$$2\pi \Delta P R^2 \, y \, dy = 2\pi L R \, y \, d\tau + 2\pi L R \, \tau \, dy \tag{7-2}$$

Integrating and defining the radius of zero shear stress by the boundary condition $\tau = 0$ at $y = \lambda$,

$$\tau = \frac{\Delta P R}{2L} \left(\frac{\lambda^2}{y} - y \right) \tag{7-3}$$

Substituting Equation (7-1),

$$\gamma = f \left[\frac{\Delta P R}{2L} \left(\frac{\lambda^2}{y} - y \right) \right] \tag{7-4}$$

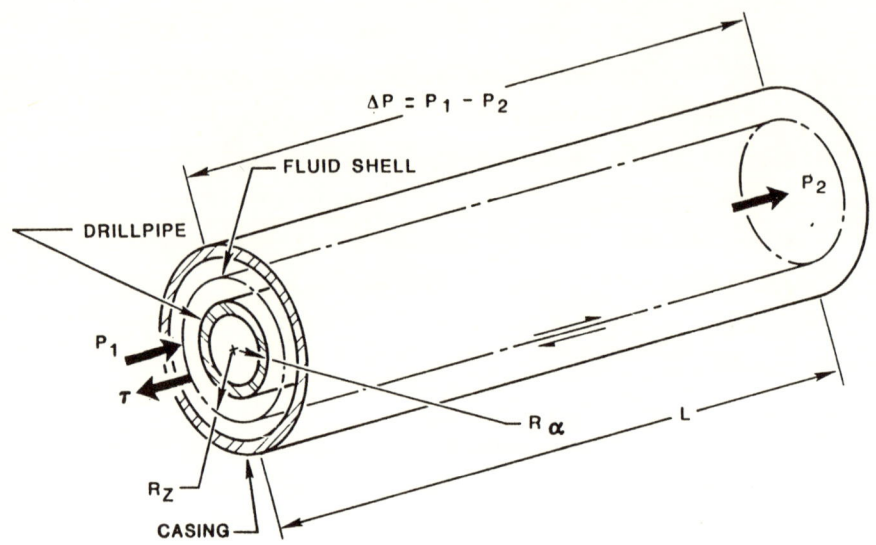

Figure 7-3. Forces in Equilibrium for Annular Pipe Flow

Limits of the plug-flow region may be found by applying Equation (7-3):

$$\tau_o = \frac{\Delta PR}{2L}\left(\frac{\lambda^2}{\lambda_-} - \lambda_-\right) \tag{7-5}$$

$$-\tau_o = \frac{\Delta PR}{2L}\left(\frac{\lambda^2}{\lambda_+} - \lambda_+\right) \tag{7-6}$$

Adding and simplifying these, we find

$$\lambda^2 = \lambda_- \lambda_+ \tag{7-7}$$

The radius corresponding to zero shear stress is therefore the geometric mean of the boundaries of the plug-flow region. Substituting (7-7) into (7-5),

$$\frac{2L\tau_o}{\Delta PR} = \lambda_+ - \lambda_- \tag{7-8}$$

The dimensionless width of the plug-flow region is thus inversely proportional to the pressure drop. All the above expressions may be applied to any fluid model, although if there is no yield stress there will be no plug-flow region and $\lambda_- = \lambda = \lambda_+$.

Velocity of the fluid may be found by integrating the expression for shear rate with a boundary condition of zero velocity at the wall. Thus,

$$v = R \int_{z=\alpha}^{z=y} f\left[\frac{\Delta PR}{2L}\left(\frac{\lambda^2}{z} - z\right)\right] * dz \qquad (7-9)$$

Velocity is known to be zero at the outer wall, so

$$R \int_{z=\alpha}^{z=1} f\left[\frac{\Delta PR}{2L}\left(\frac{\lambda^2}{z} - z\right)\right] * dz = 0 \qquad (7-10)$$

One more equation is needed to define the system, and this is obtained by integrating the velocities to obtain flowrate:

$$Q = 2\pi R^2 \int_{y=\alpha}^{y=1} vy * dy \qquad (7-11)$$

Substituting Equation (7-9):

$$Q = 2\pi R^3 \int_{y=\alpha}^{y=1} y \int_{z}^{} f\left[\frac{\Delta PR}{2L}\left(\frac{\lambda^2}{z} - z\right)\right] * dz * dy \qquad (7-12)$$

The system is defined by Equations (7-10) and (7-12). The complexity is perhaps even greater than indicated by these equations since non-Newtonian fluid models are not usually continuous functions. Therefore, the integrals must be evaluated separately over the intervals α to λ_-, λ_- to λ_+, and λ_+ to 1. Even so, it is possible to obtain analytic solutions for Newtonian and Bingham fluids (also for Power Law fluids, but only in the special case where 1/n is an integer). These solutions are as follows:

Newtonian Fluid:

$$\frac{8\mu \bar{v} L}{\Delta P R^2} = 1 + \alpha^2 + \frac{(1 - \alpha^2)}{\ln \alpha} \tag{7-13}$$

This is a single equation obtained by eliminating λ between Equations (7-10) and (7-12).

Bingham Fluid:

$$2\lambda_+ \left(\lambda_+ - \frac{2L\tau_0}{\Delta P R} \right) \ln \left(\frac{\lambda_+ - \frac{2L\tau_0}{\Delta P R}}{\lambda_+ \alpha} \right) - 1 + \frac{4L\tau_0}{\Delta P R} \left(1 - \lambda_+ \right) + \left(\frac{2L\tau_0}{\Delta P R} + \alpha \right)^2 = 0 \tag{7-14}$$

$$\frac{8\mu \bar{v} L}{\Delta P R^2} = 1 + \alpha^2 - 2\lambda_+ \left(\lambda_+ - \frac{2L\tau_0}{\Delta P R} \right) - \frac{8L\tau_0(1+\alpha^3)}{3\Delta P R(1-\alpha^2)} + \frac{2L\tau_0}{3\Delta P R(1-\alpha)^2} \left(2\lambda_+ - \frac{2L\tau_0}{\Delta P R} \right)^3 \tag{7-15}$$

Practical use of these equations requires that λ_+ be eliminated. If it is desired to find $\bar{v}$ for a known value of ΔP, the procedure is to solve Equation (7-14) iteratively for λ_+ and then substitute this value into Equation (7-15). The usual oilfield problem is to find ΔP for a known value of $\bar{v}$, in which case the above procedure must be repeated until a satisfactory solution is obtained. This can be quite a slow process, even with the aid of a computer. For this reason, and also because Equations (7-10) and (7-12) cannot be expressed analytically for Casson, Robertson-Stiff, Herschel-Bulkley, or Power Law fluids (except in the special case noted above), approximate methods are usually employed. The most useful of these is the narrow-annulus approximation, also known as the parallel-plate approximation.

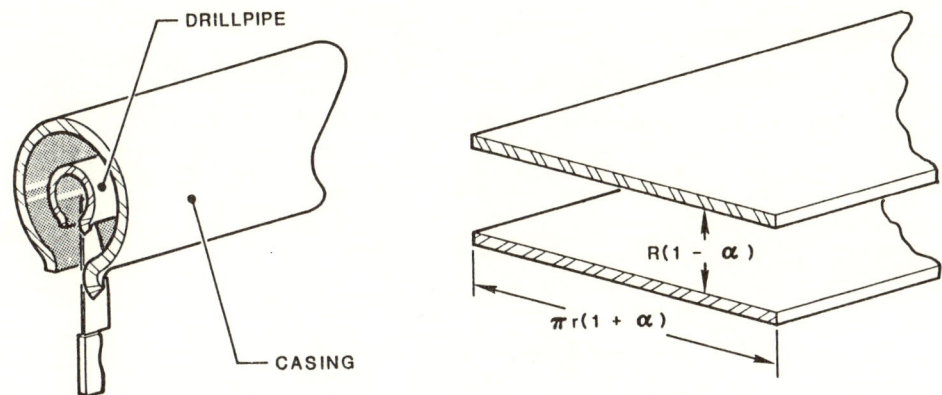

Figure 7-4. Narrow-Annulus (Parallel Plate) Model

7.4 APPROXIMATE SOLUTIONS

"Approximate" solutions are obtained by representing the annular geometry as a pair of parallel plates, of width $\pi R(1+\alpha)$, separated by a distance of $R(1-\alpha)$. This is illustrated in Figure 7-4. The parallel-plate geometry is well suited to analysis of fluid flow, and gives the following results:

Newtonian Fluid:

$$\frac{12\mu \bar{v} L}{\Delta P R^2 (1-\alpha)^2} = 1 \qquad (7-16)$$

Bingham Fluid:

$$\frac{12\mu_\infty \bar{v} L}{\Delta P R^2 (1-\alpha)^2} = 1 - \frac{3L\tau_0}{\Delta P R(1-\alpha)} + .5 \left[\frac{2L\tau_0}{\Delta P R(1-\alpha)}\right]^3 \qquad (7-17)$$

Casson Fluid:

$$\frac{12\mu_\infty \bar{v} L}{\Delta P R^2 (1-\alpha)^2} = 1 - \frac{12}{5}\left[\frac{2L\tau_o}{\Delta PR(1-\alpha)}\right]^{.5} + \frac{3L\tau_o}{\Delta PR(1-\alpha)} - \frac{1}{10}\left[\frac{2L\tau_o}{\Delta PR(1-\alpha)}\right]^3 \quad (7\text{-}18)$$

Power Law Fluid:

$$\frac{6\bar{v}}{R(1-\alpha)}\left[\frac{2kL}{\Delta PR(1-\alpha)}\right]^{1/n} = \frac{3n}{(2n+1)} \quad (7\text{-}19)$$

Robertson-Stiff Fluid:

$$\frac{12\bar{v}Lk\,\gamma_o^n}{\Delta PR^2(1-\alpha)^2 \gamma_o} = \frac{3n}{(2n+1)}\left[\frac{\Delta PR(1-\alpha)}{2Lk\gamma_o^n}\right]^{(\frac{1}{n}-1)} - \frac{3Lk\gamma_o^n}{\Delta PF(1-\alpha)}$$

$$+ \frac{3}{2(2n+1)}\left[\frac{2Lk\gamma_o^n}{\Delta PR(1-\alpha)}\right]^3 \quad (7\text{-}20)$$

It is not possible to obtain an analytic solution for Herschel-Bulkley fluids.

The error introduced by the parallel-plate approximation depends on the annulus diameter ratio α and on the degree to which the fluid departs from Newtonian behavior. For a Power Law fluid the degree of non-Newtonian behavior is characterized by the flow behavior index n. If n = 1, the fluid is Newtonian, and if n < 1, the fluid is pseudoplastic or shear-thinning. The percentage error in calculated pressure drop caused by the parallel-plate approximation is shown in Figure 7-5.

Two interesting points arise from this table. First, as long as the annulus diameter ratio is greater than 0.2 (as is nearly always the case in oilfield calculations), the error is less than 4 percent. This is perfectly acceptable, being less than the precision to which viscometer readings are taken. Second, increasingly non-Newtonian (pseudoplastic) behavior actually reduces the error in the calculation.

Diameter Ratio →	0.0	0.1	0.2	0.3	0.4	0.5	0.6	0.7	0.8	0.9	1.0
n ↓											
1.0	50.00	7.42	3.95	2.30	1.36	0.79	0.43	0.21	0.08	0.02	0.00
0.5	26.49	6.77	3.64	2.14	1.27	0.74	0.40	0.19	0.08	0.02	0.00
0.333	18.56	5.98	3.29	1.92	1.16	0.67	0.34	0.18	0.08	0.02	0.00
0.25	14.42	5.27	2.93	1.75	1.04	0.61	0.33	0.15	0.07	0.02	0.00
0.2	11.84	4.71	2.65	1.58	0.96	0.56	0.32	0.16	0.07	0.02	0.00

Figure 7-5. Narrow-Annulus Approximation, Power Law Fluid Percentage Error in Calculated Pressure Drop

Diameter Ratio →	0.0	0.1	0.2	0.3	0.4	0.5	0.6	0.7	0.8	0.9	1.0
$\dfrac{2\tau_o R(1-\alpha)}{\mu_\infty \overline{V}}$ ↓											
0	50.00	7.42	3.95	2.30	1.36	0.79	0.43	0.21	0.08	0.02	0.00
1	44.62	7.02	3.75	2.18	1.29	0.75	0.41	0.20	0.08	0.02	0.00
2	40.52	6.71	3.59	2.10	1.24	0.72	0.39	0.19	0.08	0.02	0.00
5	32.42	6.05	3.26	1.90	1.12	0.66	0.37	0.18	0.07	0.02	0.00
10	25.29	5.36	2.91	1.71	1.01	0.59	0.32	0.17	0.07	0.01	0.00
20	18.72	4.54	2.49	1.46	0.87	0.51	0.28	0.14	0.05	0.01	0.00
50	12.04	3.40	1.89	1.12	0.67	0.40	0.22	0.11	0.04	0.00	0.00
100	8.50	2.62	1.48	0.88	0.54	0.31	0.18	0.08	0.03	0.00	0.00
200	5.99	1.97	1.12	0.67	0.41	0.24	0.13	0.06	0.03	0.00	0.00
500	3.76	1.30	0.75	0.45	0.28	0.16	0.09	0.04	0.02	0.00	0.00
1000	2.65	0.95	0.55	0.33	0.20	0.11	0.07	0.03	0.01	0.00	0.00

Figure 7-6. Narrow-Annulus Approximation, Bingham Fluid Percentage Error in Calculated Pressure Drop

A similar table for Bingham fluids is presented in Figure 7-6 (page 7-9). The degree of non-Newtonian behavior is characterized by the dimensionless group

$$\frac{2\tau_0 R(1-\alpha)}{\mu_\infty \overline{V}}$$

which is sometimes called the plasticity (or the Bingham) number. Again, the error is acceptable if the diameter ratio exceeds 0.2, and increasingly non-Newtonian fluids give smaller errors.

Based on the behavior of Power Law and Bingham fluids, one would intuitively expect that the narrow-annulus approximation is similarly valid for Casson and Robertson-Stiff fluids since both represent behavior intermediate between Bingham and Power Law models. However, this has not been rigorously proven.

7.5 PRACTICAL METHODS

7.6 Newtonian Fluid

There is no difficulty in analyzing annular flow of a Newtonian fluid. An exact solution may be obtained by use of Equation (7-13), or the narrow-annulus approximation may be used as in Equation (7-16).

7.7 Bingham Fluid

The exact solution as represented by Equations (7-14) and (7-15) can be found with the aid of a computer, but the process is tedious and may be too slow for real-time applications. The parallel-plate approximation gives excellent results, and Equation (7-17) may be solved as follows:

$$B = \frac{2\tau_0 R(1-\alpha)}{\mu_\infty \overline{V}} \qquad (7-21)$$

$$\beta = \left(\frac{B}{B+8}\right)^{.5} \qquad (7-22)$$

$$x = \frac{2}{\beta} \sin\left[\frac{1}{3} \sin^{-1}\left(\beta^3\right)\right] \qquad (7-23)$$

$$\Delta P = \frac{2L\tau_0}{xR(1-\alpha)} \qquad (7-24)$$

This method is convenient for implementation on programmable calculators or small computer systems. It is still rather inconvenient for manual calculations, and a popular oilfield approximation has been to omit the cubic term from Equation (7-17). The resulting expression can be rearranged as

$$\Delta P \approx \frac{12\mu_\infty \bar{v} L}{R^2(1-\alpha)^2} + \frac{3\tau_0 L}{R(1-\alpha)} \qquad (7-25)$$

or, as commonly seen in oilfield units

$$\Delta P = \frac{PV * V * L}{60,000(d_h - d_p)^2} + \frac{YP * L}{200(d_h - d_p)}$$

where
- L = section length (ft)
- YP = yield point (lb/100 ft^2)
- $(d_h - d_p)$ = hole diameter minus pipe outside diameter (inches)
- PV = plastic viscosity
- V = annular velocity in laminar flow (ft/min)

This expression is very much less accurate than Equation (7-17), and the solution given in Equations (7-21) through (7-24) is to be preferred when possible. Equation (7-25) gives acceptable results when the plug-flow region is small (low yield point or high velocity), but the errors can reach 50 percent at other times. Figure 7-7 shows the magnitude of error introduced by the approximation.

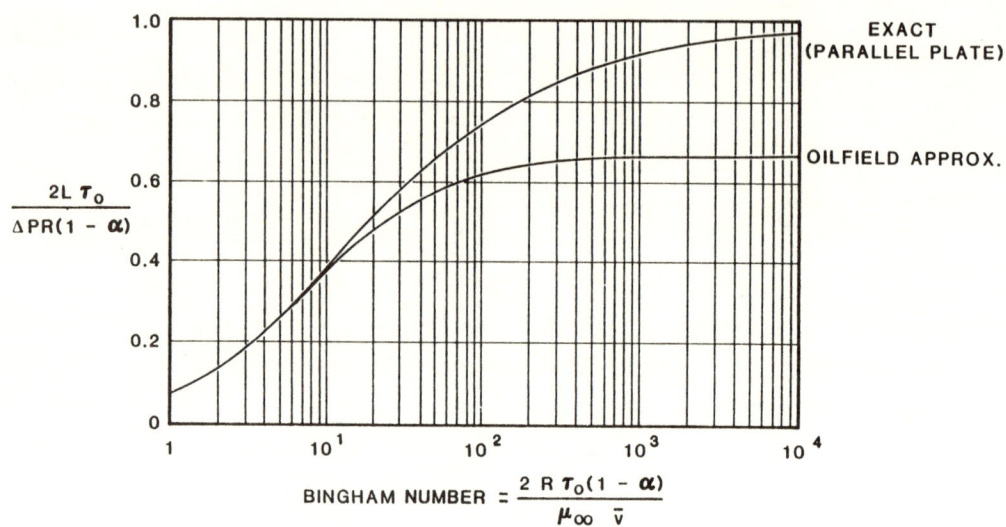

Figure 7-7. Approximation Error for Annular Flow of Bingham Fluid

7.8 Power Law Fluid

The narrow-annulus solution of Equation (7-19) is convenient and widely used. A further development makes use of the exact solution to Equations (7-10) and (7-12), using an empirical correlation between dimensionless parameters. This method gives results accurate to 1 percent over all annular geometries. The correlation is as follows:

$$y = .37n^{-.14} \tag{7-26}$$

$$z = 1 - (1 - \alpha^y)^{1/y} \tag{7-27}$$

$$G = \left(1 + \frac{z}{2}\right)\left[\frac{(3-z)n+1}{(4-z)n}\right] \tag{7-28}$$

$$\Delta P = \frac{2kL}{R(1-\alpha)} \ast \left[\frac{4\bar{v}G}{R(1-\alpha)}\right]^n \tag{7-29}$$

This method is often used when a computer system is available. It does give better results than the narrow-annulus approximation, although the improvement is slight for annulus diameter ratios exceeding 0.3.

7.9 Casson Fluid

The narrow-annulus formula of Equation (7-18) may be expected to give good results. It requires an iterative solution and therefore may be too time consuming for many applications.

7.10 Robertson-Stiff Fluid

The only practical way to obtain results is to use the narrow-annulus approximation, Equation (7-20). This must be solved iteratively.

7.11 TURBULENT FLOW

As in the case of pipe flow, the degree of turbulence can be characterized by the Reynolds number. It is necessary to define the equivalent diameter to be used in the calculation of the Reynolds number; usually, the hydraulic diameter is used. Hydraulic diameter is defined as four times the cross-sectional area of the flow channel, divided by its wetted perimeter. For a circular annulus this becomes simply the difference between the outer and inner diameters.

$$Re = \frac{\rho \bar{v} D (1-\alpha)}{\bar{\mu}} \tag{7-30}$$

where $\alpha = d/D$. The mean viscosity, $\bar{\mu}$, is the viscosity of the Newtonian fluid which in laminar flow would develop the same pressure drop. From Equation (7-13),

$$\bar{\mu} = \frac{\Delta P_L R^2}{8L\bar{v}} \left[1 + \alpha^2 + \frac{(1-\alpha^2)}{\ln \alpha} \right] \tag{7-31}$$

In practice, it is unlikely that exact annulus-flow solutions will be required except for Newtonian or Power Law fluids. For other fluid types, the narrow-annulus approximation will probably be used. In this case, from Equation (7-16) the expression for mean viscosity becomes

$$\bar{\mu} = \frac{\Delta P_L \, R^2 (1 - \alpha)^2}{12 \, L \, \bar{v}} \qquad (7\text{-}32)$$

It has been demonstrated experimentally that for Newtonian fluids the turbulent-flow relationship between friction factor and Reynolds number remains valid, whether flow is in circular pipes or between parallel plates. Since these represent the two extreme cases of annular geometry, it is reasonable to assume that the same relationship may be used for any annular geometry.

Little experimental data is available on turbulent flow of non-Newtonian fluids through annuli. Until evidence otherwise is provided, it must be assumed that the friction factor/Reynolds number relationship for pipe flow applies equally well to other geometries.

In summary, pressure losses in the annulus for turbulent flow may be calculated in nearly the same manner as for turbulent flow in the pipe. The Fanning equation is used

$$\Delta P = \frac{2 f L \rho \bar{v}^2}{D}$$

with only small modifications due to the annular geometry. The hydraulic diameter (D-d for a circular annulus) replaces D in both the Fanning equation and the Reynolds number. A different mean viscosity, defined for an annulus by Equations (7-31) and (7-32), is also used in the Reynolds number. Once the Reynolds number has been calculated for annular flow, the mathematical expressions and graphed curves presented in under paragraph heading 5.8 may be used for calculating the friction factor f.

7.12 TRANSITIONAL FLOW

Applying the criterion of transition at a laminar friction factor of .008, a Newtonian fluid will enter turbulent flow at a Reynolds number of 3000. The critical Reynolds number is 50 percent higher in parallel-plate flow than in pipe flow. Experiments have shown, at least for Newtonian fluids, that the transition does occur close to this point.

For Bingham fluids, data is provided again by the work of Hanks and Pratt. The following calculations are consistent with their work, although the equations have been modified to account for different definitions of both the Hedstrom number and the Reynolds number.

$$He = \frac{\rho D^2 (1 - \alpha)^2}{\mu_\infty^2} \tau_o \qquad (7\text{-}33)$$

$$33,600 \, x_c = He(1 - x_c)^3 \quad [\text{solve for } x_c] \qquad (7\text{-}34)$$

$$Re_c = \frac{He}{12 x_c} \left[1 - \frac{3 x_c}{2} + \frac{x_c^3}{2} \right]^2 \qquad (7\text{-}35)$$

The limiting value of Re_c is 2800 for a Newtonian fluid (as $He \to 0$), as illustrated in Figure 7-8.

Substituting either Equation (7-31) or (7-32) for the mean viscosity in Equation (7-30) and solving for velocity at the critical Reynolds number will yield a critical velocity. A common expression for the critical velocity of a Bingham fluid uses Equations (7-25) and (7-32), and considers turbulence to occur at a critical Reynolds number of 2000. This expression is

$$v_c = \frac{1000 \, \mu_\infty + 1000 \sqrt{\mu_\infty^2 + \dfrac{\rho \tau_o D^2 (1-\alpha)^2}{4000}}}{\rho D (1-\alpha)} \qquad (7\text{-}36)$$

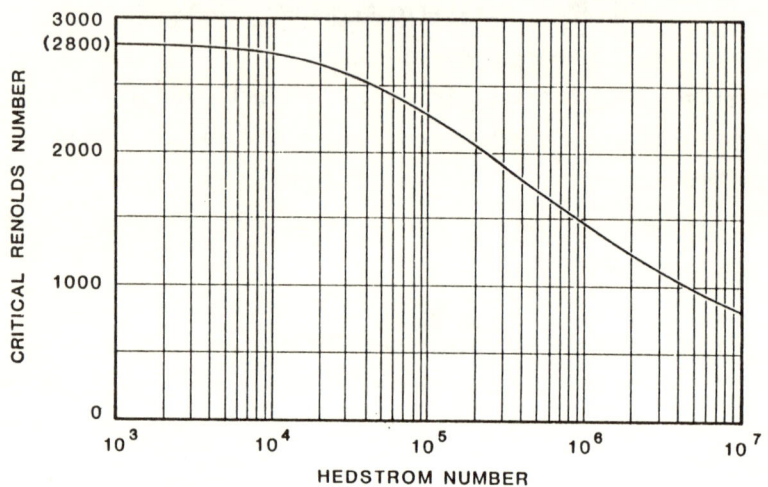

Figure 7-8. Critical Reynolds Number for Annular Flow, by Method of Hanks and Pratt

7.13 REFERENCES

1. Bootwala, I., "Method Speeds Drill Site Hydraulics Calculation," World Oil, April 1979.

2. Carlton, L. A., and M. E. Chenevert, "A New Approach to Preventing Lost Returns," S.P.E. Paper 4972, 1974.

3. Coleman, B. D., and W. Noll, "Helical Flow of General Fluids," J. App. Phys. _30_ (10), 1959.

4. Fredrickson, A. G., "Helical Flow of an Annular Mass of Visco-Elastic Fluid," Chem. Eng. Sci. _11_, 1960.

5. Fredrickson, A. G., and R. B. Bird, "Non-Newtonian Flow in Annuli," Ind. and Eng. Chem. _50_ (3), 1958.

6. Hanks, R. W., and D. R. Pratt, "On the Flow of Bingham Plastic Slurries in Pipes and Between Parallel Plates," S.P.E. J., December 1967.

7. Hanks, R. W., and M. P. Valia, "A Theory of Transitional and Turbulent Flow of non-Newtonian Slurries Between Flat Parallel Plates," S.P.E. J., March 1971.

8. Iyoho, A. W., "An Accurate Slot-Flow Model for non-Newtonian Fluid Flow through Eccentric Annuli," S.P.E. Paper 9447, 1980.

9. Melrose, J. C., J. G. Savins, W. R. Foster and E. R. Parish, "A Practical Utilization of the Theory of Bingham Plastic Flow in Stationary Pipes and Annuli," Pet. Trans. A.I.M.E. 213, 1958.

10. Metzner, A. B., and J. C. Reed, "Flow of non-Newtonian Fluids -- Correlation of the Laminar, Transition, and Turbulent-Flow Regions," A.I.Ch.E. J., 1 (4), December 1955.

11. Moore, P. L., Drilling Practices Manual, Petroleum Publishing Co., 1974.

12. Paslay, P. R., and A. Slibar, "Laminar Flow of Drilling Mud Due to Axial Pressure Gradient and External Torque," Pet. Trans. A.I.M.E., 210, 1957.

13. Randall, B. V., and D. B. Anderson, "Flow of Mud During Drilling Operations," S.P.E. Paper 9444, 1980.

14. Rotem, Z., "Non-Newtonian Flow in Annuli," J. App. Mech., June 1962.

15. Savins, J. G., "Generalized Newtonian (Pseudoplastic) Flow in Stationary Pipes and Annuli," Pet. Trans. A.I.M.E. 213, 1958.

16. Savins, J. G., and G. C. Wallick, "Viscosity Profiles, Discharge Rates, Pressures, and Torques for a Rheologically Complex Fluid in a Helical Flow," A.I.Ch.E. J. 12(a), 1966.

17. Schuh, F. J., "Computer Makes Surge Pressure Calculations Useful," O&G J., August 3, 1964.

18. Siegel, B., "How to Calculate Annular Pressure Loss and Critical Velocity," Pet. Engr., September 1980.

19. Taylor, R., and D. Smalling, "A New and Practical Application of Annular Hydraulics," S.P.E. Paper 4518, 1973.

20. Vaughn, R. D., "Axial Laminar Flow of non-Newtonian Fluids in Concentric Annuli," S.P.E. J., December 1963.

21. Vaughn, R. D., "Axial Laminar Flow of non-Newtonian Fluids in Narrow Eccentric Annuli," S.P.E. J., December 1965.

22. Walker, R. E., "Migration of Particles to a Hole Wall in a Drilng Well," S.P.E. J., June 1969.

23. Walker, R. E., and O. Al-Rawi, "Helical Flow of Bentonite Slurries," S.P.E. Paper 3108, 1970.

24. Walker, R. E., and D. E. Korry, "Field Method of Evaluating Annular Performance of Drilling Fluids," S.P.E. Paper 4321, 1973.

25. Wallick, G. C., and J. G. Savins, "A Comparison of Diferential and Integral Descriptions of the Annular Flow of a Power-Law Fluid," S.P.E. J., September 1969.

26. Wang, Z. Y., and S. R. Tang, "Casson Rheological Model in Drilling Fluid Mechanics," S.P.E. Paper 10564, 1982.

27. Whan, G. A., and R. R. Rothfus, "Characteristics of Transition Flow Between Parallel Plates," A.I.Ch.E. J. $\underline{5}$ (2), June 1959.

28. Wheeler, J. A., and E. H. Wissler, "The Friction Factor-Reynolds Number Relation for the Steady Flow of Pseudoplastic Fluids through Rectangular Ducts," A.I.Ch.E. J. 11, March 1965.

29. Wilson, L. E., "Results of a Field Test on Circulating and Surge Pressures," World Oil, September 1962.

30. Zamora, M., and D. L. Lord, "Practical Analysis of Drilling Mud Flow in Pipes and Annuli," S.P.E. Paper 4976, 1974.

8
SWAB AND SURGE

8.1   GENERAL

Swab and surge pressures, caused by moving the drillstring axially, can be calculated by a method similar to that used for calculating annular circulating pressures (Section 7). The greatest problem is determining fluid flowrate in the annulus when the pipe is open-ended since the distribution of flow between pipebore and annulus cannot be found by any simple method.

Two approaches to this problem have been proposed. The first assumes that fluid levels in the annulus and pipebore are equal at all times. The distribution of fluid displaced by the drillstring is therefore dependent on the relative cross-sectional areas of annulus and pipebore, as shown in Figure 8-1. The simple expression for annular velocity is given by

$$\bar{v} = -v_p \frac{(d^2 - d_i^2)}{(D^2 - d^2 + d_i^2)} \tag{8-1}$$

where

$\bar{v}$ = average velocity
v_p = pipe velocity
D = outer diameter
d = outside diameter of drillstring
d_i = inside diameter of drillstring

The minus sign occurs because the pipe velocity is in the opposite direction of the fluid velocity.

Equation (8-1) remains valid even when the hole geometry changes. This method is easy to apply and is in widespread use. But its basic premise, that fluid

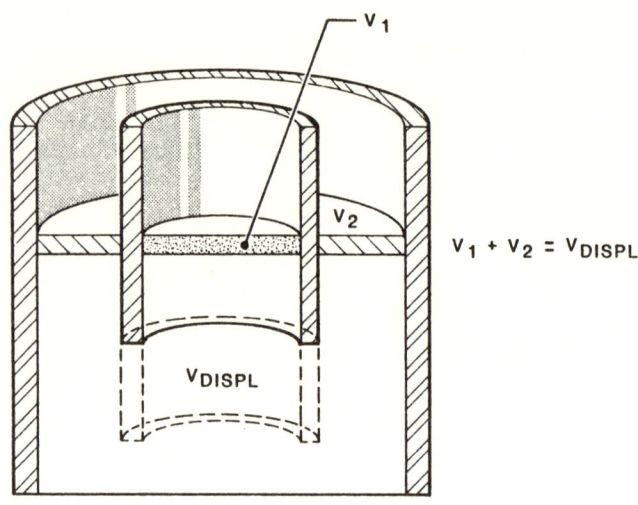

Figure 8-1. Equal Level Displacement

levels in drillpipe and annulus remain equal, is rarely justifiable. Because of the greater restriction to flow presented by the bit nozzles and pipebore, actual flow in the annulus will nearly always exceed that calculated by this method. Calculated swab and surge pressures will therefore usually be too low.

An alternative procedure considers the pipebore and annulus as a U-tube, as shown in Figure 8-2. It is clear that the sum of hydrostatic and frictional pressures in the pipebore and through the bit should equal the sum of hydrostatic and frictional pressures in the annulus. Both sums represent the pressure prevailing immediately below the bit. There is only one flow distribution which will fulfill this criterion, and it can be found by trial and error through the use of the pressure loss equations and a computer.

When pulling out of the hole it may be assumed that both pipe and annulus are kept full of fluid. The required distribution of flow is therefore that which gives equal frictional losses in pipebore and annulus. When running in the hole, however, the fluid level inside the drillstring may drop well below that in the annulus if small bit nozzles are present. This effect can frequently be observed as pit volume higher than expected, string weight less than expected, and a considerable volume pumped before standpipe pressure builds up

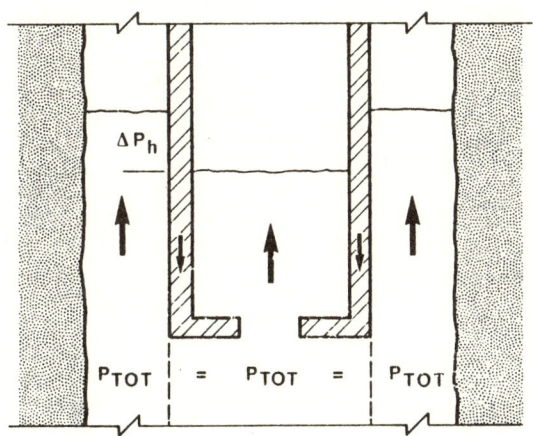

Figure 8-2. U-Tube Analogy for Equal Pressure Displacement

while breaking circulation. When the fluid level in the drillstring is below that of the annulus, a greater hydrostatic pressure will exist in the annulus, and fluid will tend to flow from the annulus up the drillstring. In this case, calculating flow distribution by equating frictional losses gives a calculated annular flow and surge pressure slightly higher than actually exists. Because the error is small and conservative, and because at present there is no reliable way of measuring the fluid level within the pipe, the practice of calculating flow distribution by equating internal and external frictional pressure losses is generally acceptable.

When the pipe is closed or contains a float sub, it is easy to calculate flow in the annulus. All of the fluid displaced by the drillstring passes up the annulus, so

$$\bar{v} = - v_p * \frac{d^2}{(D^2 - d^2)} \qquad (8\text{-}2)$$

Calculating the pressure drop in the annulus is complicated by the motion of the inner wall. This motion is in the opposite direction to the displaced fluid, so pressure drop will be greater than for the same flowrate in a stationary annulus. Equations describing the system can be formulated, but their solutions are generally too tedious for wellsite use.

The problem can be solved for a Newtonian fluid in laminar flow. The solution can be expressed as

$$\frac{8\mu L}{\Delta P R^2}\left[\bar{v} + \frac{v_p}{2 \ln \alpha} + \frac{v_p * \alpha^2}{(1-\alpha^2)}\right] = 1 + \alpha^2 + \frac{(1-\alpha^2)}{\ln \alpha} \qquad (8\text{-}3)$$

where
v_p = pipe velocity
α = d/D

The analogy with Equation (7-13), the stationary annulus solution, is clear. The stationary annulus solution can be used if an effective fluid velocity, is substituted:

$$v_{eff} = \bar{v} + v_p\left[\frac{1}{2 \ln \alpha} + \frac{\alpha^2}{(1-\alpha^2)}\right] \qquad (8\text{-}4)$$

The term

$$\frac{1}{2 \ln \alpha} + \frac{\alpha^2}{(1-\alpha^2)}$$

is known as the clinging constant, K. It represents the proportion of pipe velocity which must be added to fluid velocity in order to find the equivalent or effective velocity which can be used in the stationary annulus calculation. The effective velocity is numerically greater than the actual fluid velocity: because $\bar{v}$ and v_p are of opposite sign, the clinging constant is negative.

For Newtonian fluids, the clinging constant depends only on the annulus diameter ratio. This is not the case for non-Newtonian fluids, in which the clinging constant is also a function of the fluid's properties and of fluid and pipe velocity. Calculation of clinging constants for non-Newtonian fluids is very tedious: some representative values for Bingham fluids and Power Law are shown in Figures 8-3 and 8-4. For Figure 8-3, B_p is the Bingham number in oilfield units, where $\bar{v}$ = pipe velocity = v_p.

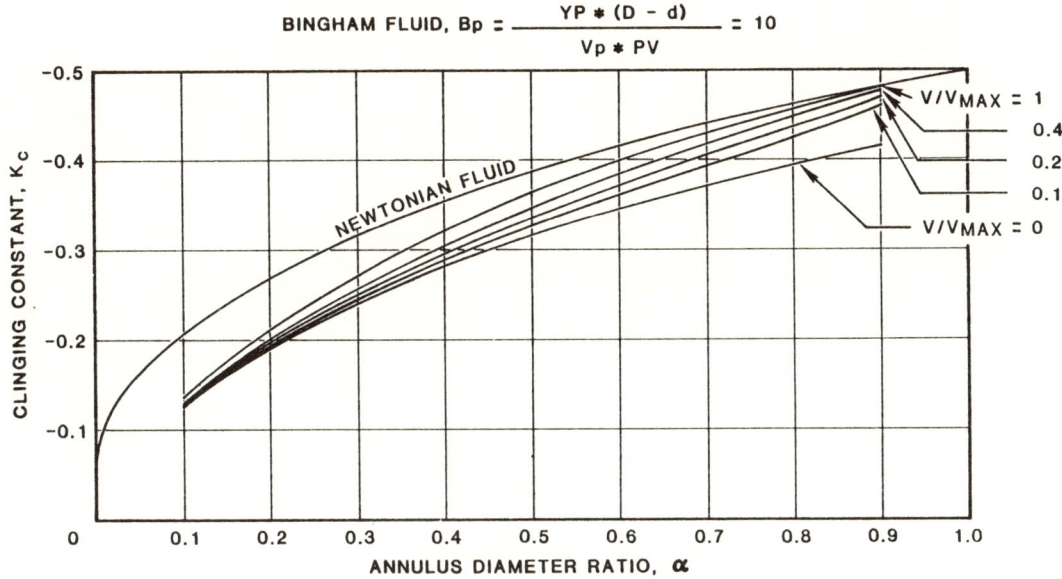

Figure 8-3. Clinging Constants for Bingham Fluid

In the oilfield it is a common practice to assume a clinging constant of -0.45; but Figures 8-3 and 8-4 show that this can be considerably in error, and a more exact calculation is preferred.

The possibility of turbulent flow in the annulus must be considered in swab and surge pressure calculations. Assuming that the velocity profile in turbulent flow is flat, the resultant pressure drop can be estimated by summing the pressure drops caused by velocity $\overline{v}$ at the outer wall and the relative velocity ($\overline{v} - v_p$) at the inner wall, because $\overline{v}$ and v_p are of opposite sign. With closed pipe, the turbulent flow clinging constant can be approximated by

$$K_t \simeq \frac{\alpha^2 - \left[\dfrac{\alpha^4 + \alpha}{1 + \alpha}\right]^{.5}}{1 - \alpha^2} \qquad (8\text{-}5)$$

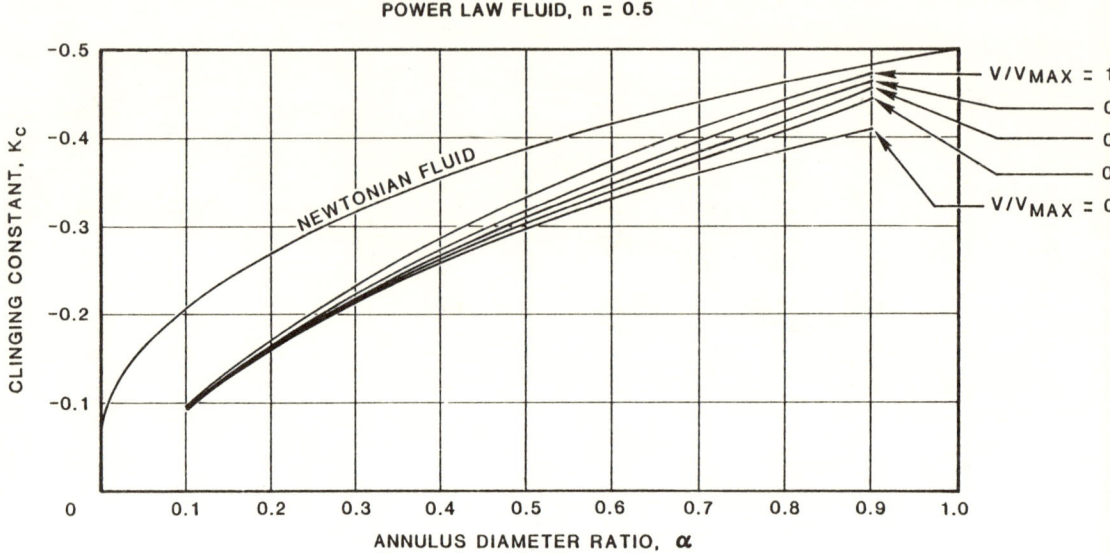

Figure 8-4. Clinging Constant for Turbulent Flow

This function is graphed in Figure 8-5. It is an approximation for closed-ended pipe and it appears to be similar to turbulent flow clinging constants derived by other authors.

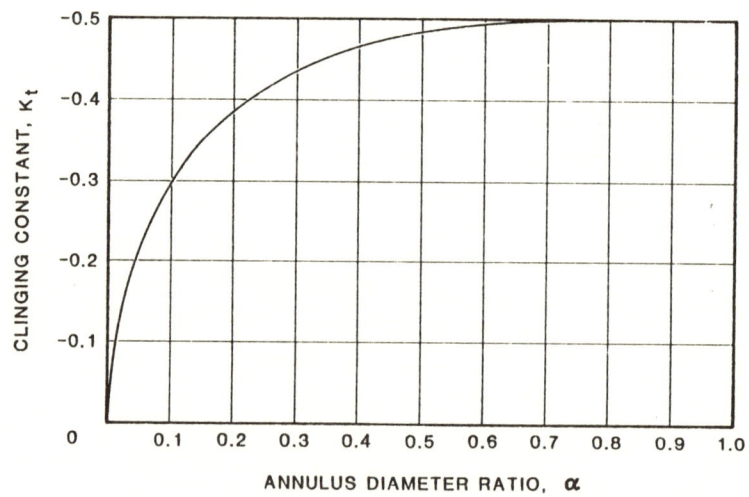

Figure 8-5. Clinging Constant for Turbulent Flow

The point at which transition occurs from laminar to turbulent flow is difficult to determine theoretically, and experimental data is lacking. It is therefore recommended that, for swab or surge pressure calculations, both laminar and turbulent pressure drops should be calculated. The flow regime giving the greater pressure drop may then be considered to be correct. This procedure is conservative in that it will give a pressure equal to or exceeding the true swab or surge pressure.

Finally, when calculating swab or surge pressures for Power Law fluids, the calculated frictional pressure loss should be checked against the pressure required to break gels. For each section of the annulus, the gel-breaking pressure is

$$P_g = \frac{4L\tau_g}{(D - d)} \qquad (8\text{-}6)$$

where

P_g = gel-breaking pressure
L = section length
τ_g = gel strength
D = outer diameter
d = inner diameter

If the calculated swab or surge pressure is less than the sum of the gel-breaking pressures, the gel-breaking pressure should be used. This check is not required if the fluid model incorporates a yield stress.

8.2 REFERENCES

1. Burkhardt, J. A., "Wellbore Pressure Surges Produced by Pipe Movement," J.P.T., June 1961.

2. Cardwell, W. T., "Pressure Changes in Drilling Wells Caused by Pipe Movement," Drilling and Production Practices, A.P.I., 1953.

3. Carlton, L. A., and M. E. Chenevert, "A New Approach to Preventing Lost Returns," S.P.E. Paper 4972, 1974.

4. Fontenot, J. E., and R. K. Clark, "An Improved Method for Calculating Swab and Surge Pressures in a Drilling Well," S.P.E. J., October 1964.

5. Moore, P. L., <u>Drilling Practices Manual</u>, Petroleum Publishing Co., 1974.

6. Robinson, L. H., J. M. Speers, L. A. Watkins, A. Barry and J. F. Miller, "Exxon MWD Tools Yield Unexpected Downhole Data," O&G J., April 21, 1980.

7. Schuh, F. J., "Computer Makes Surge Pressure Calculations Useful," O&G J., August 3, 1964.

8. Wilson, L. E., "Results of a Field Test on Circulating and Surge Pressures," World Oil, September 1962.

9
CUTTINGS TRANSPORT

9.1 GENERAL

One of the primary functions of a drilling fluid is to bring drilled cuttings to the surface. Inadequate hole cleaning can lead to a number of problems, including fill, packing off, stuck pipe, and excessive hydrostatic pressure.

The ability of a drilling fluid to lift cuttings is affected by many factors, and there is no universally-accepted theory which can account for all observed phenomena. Some of these parameters are fluid density and rheology, annulus size and eccentricity, annular velocity and flow regime, pipe rotation, cuttings density, and size and shape of the cuttings.

Figure 9-1 shows some general features of cuttings transport in laminar flow. If the particles are of irregular shape, they are subjected to a torque caused by the shearing of the mud. If the drillpipe is rotating, a centrifugal effect causes particles to move toward the outer wall of the annulus. The process is further complicated because the viscosity of non-Newtonian fluids varies according to the shear rate, and therefore the viscosity changes with radial position. Finally, transport rates are strongly dependent on particle size and shape, which, in a field situation, are both irregular and variable.

The only practical way to estimate the slip velocity (or relative sinking velocity) of cuttings is to develop empirical correlations based on experimental data. Even with this approach, there is quite a wide disparity in the results obtained by different authors.

It has been proposed that the correlation can be expressed as a relationship between a particle Reynolds number, which represents the geometry of flow around the particles, and a drag coefficient. The particle Reynolds number for a spherical particle (or disk) is usually defined as

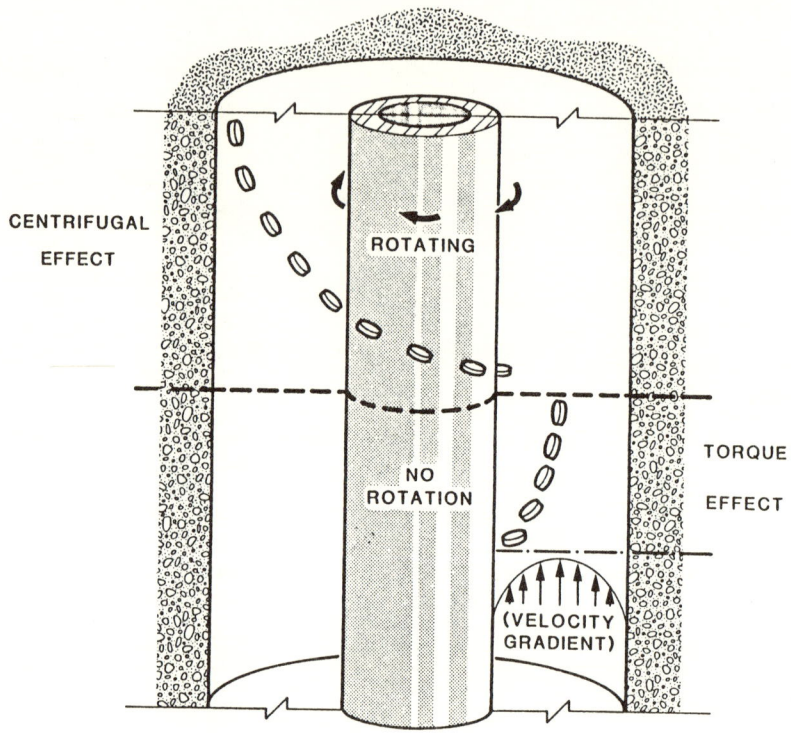

Figure 9-1. Cuttings Transport

$$Re_p = \frac{\rho_f * v_s * d_p}{\bar{\mu}} \qquad (9-1)$$

where

- Re_p = particle Reynolds number
- ρ_f = fluid density
- v_s = particle slip velocity
- d_p = particle diameter
- $\bar{\mu}$ = mean viscosity (as defined in under paragraph heading 5.6)

This definition has the advantage that it can be applied to any fluid model as long as the pressure drop in annular flow can be calculated or measured. Some authors have used alternative definitions, however.

The drag coefficient, also called friction factor by some authors, can be calculated for a spherical particle as

$$C_D = \frac{4 d_p (\rho_p - \rho_f) g}{3 \rho_f v_s^2} \qquad (9\text{-}2)$$

In the case of a disk settling flatwise, this becomes

$$C_D = \frac{2 h (\rho_p - \rho_f) g}{\rho_f v_s^2} \qquad (9\text{-}3)$$

where

    h = thickness of disk

Experimental data can be correlated as a relationship between particle Reynolds number and drag coefficient. Figures 9-2 and 9-3 show the data of Walker and Mayes and of Moore, together with their proposed correlation curves. Particle slip velocity can be calculated using Equation (9-1), using the appropriate drag coefficient, and inverting Equations (9-2) or (9-3) to solve for slip velocity.

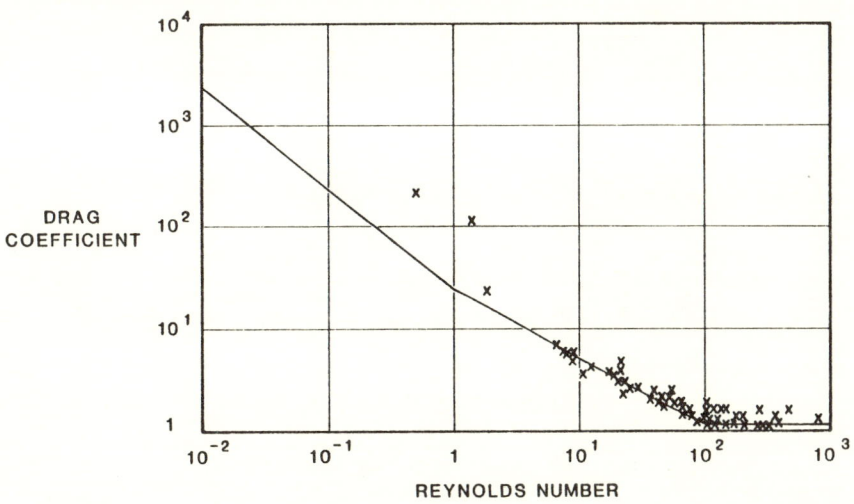

Figure 9-2. Walker and Mayes' Correlation for Drag Coefficient

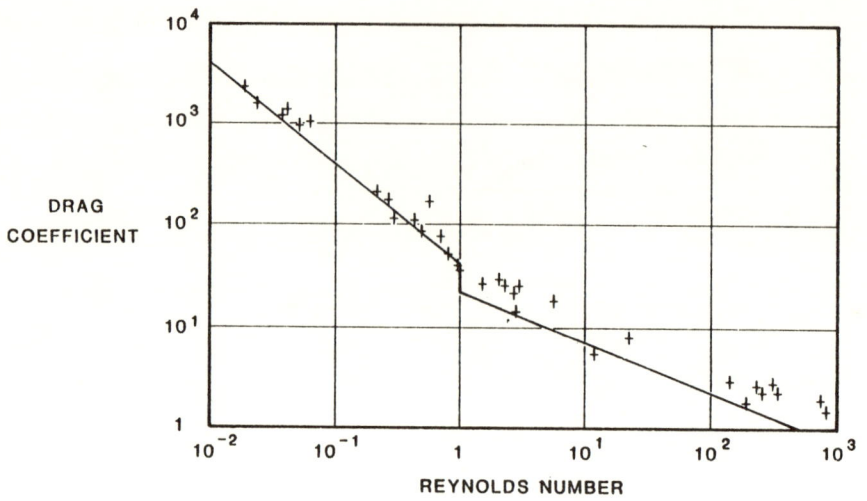

Figure 9-3. Moore's Correlation for Drag Coefficient

An evaluation of several correlations was undertaken by Sample and Bourgoyne. Although no one method gave consistently good results, Moore's correlation gave the smallest average error. This correlation can be expressed as

$$C_D = \frac{40}{\sqrt{Re_p}} \qquad (Re_p \leqslant 1) \qquad (9\text{-}4)$$

$$C_D = \frac{22}{\sqrt{Re_p}} \qquad (1 < Re_p < 2000) \qquad (9\text{-}5)$$

$$C_D = 1.5 \qquad (2000 \leqslant Re_p) \qquad (9\text{-}6)$$

Although not evident in the literature, a correlation of the form

$$C_D = \frac{40}{Re_p} + 1 \qquad (9\text{-}7)$$

also fits Moore's data well, with the advantage that it is a continuous function. Only one equation is required to evaluate slip velocity, regardless of particle Reynolds number. Figure 9-4 is a combination of Figures 9-2 and 9-3,

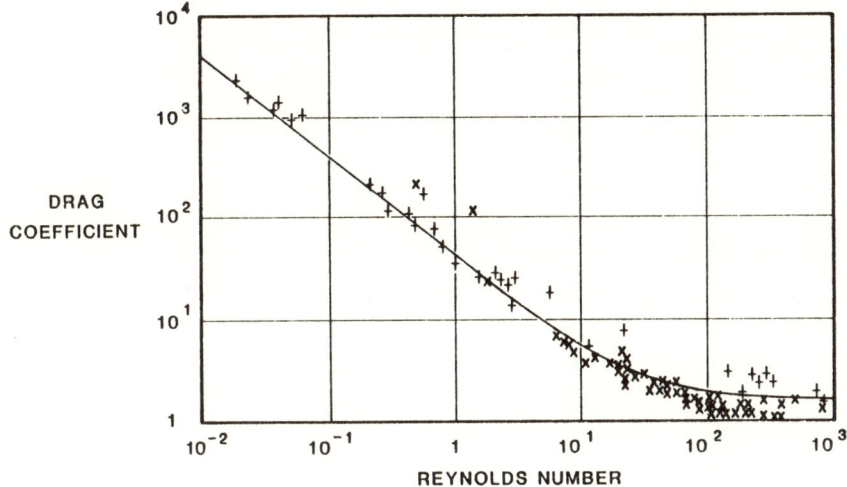

Figure 9-4. Proposed Correlation for Drag Coefficient

with the correlation proposed in Equation (9-7). The marked offset between the data of Walker and Mayes and that of Moore may result from their differing definitions of $\bar{\mu}$ and Re_p.

Equations (9-1), (9-2) and (9-7) may be combined to give a quadratic equation which can be solved for slip velocity. Thus, for spherical particles:

$$3v_s \, (40 \, \bar{\mu} + \rho_f \, v_s \, d_p) = 4g \, d_p^2 \, (\rho_p - \rho_f) \tag{9-8}$$

For flat disks falling flatwise, Equation (9-2) is replaced by (9-3), giving

$$v_s \, (40 \, \bar{\mu} + \rho_f \, v_s \, d_p) = 2g \, h \, d_p \, (\rho_p - \rho_f) \tag{9-9}$$

With this method of determining slip velocity, it is possible to calculate the concentration of cuttings in the annulus. This is a function of the cuttings feedrate, mud flowrate, and the ratio of transport velocity to annular velocity. This ratio has been called the transport ratio:

$$T = 1 - \frac{v_s}{v_m} \qquad (9\text{-}10)$$

where

T = transport ratio
v_s = slip velocity
v_m = annular velocity

When the cuttings in the annulus are at an equilibrium concentration, the volume of cuttings flowing out the annulus equals the volume of cuttings being liberated at the bit. Thus

Volume of cuttings entering annulus per unit time = Volume of cuttings leaving annulus per unit time

$$\frac{ROP * \pi D_b^2}{4} = \text{volume of cuttings leaving annulus per unit time}$$

where

D_b = bit diameter

The volume of cuttings leaving the annulus per unit time ($Q_{cuttings}$) depends upon the net velocity of the cuttings, the concentration of the cuttings and the annular area. The cuttings velocity is given by

$$v_L = v_m - v_s \qquad (9\text{-}11)$$

where

v_L = net cuttings velocity
v_m = mud velocity
v_s = slip velocity

The concentration of cuttings in the mud is denoted by C_a (0 to 1.0) and the area through which the cuttings leave the annulus is $\pi/4(D^2 - d^2)$. Thus

$$Q_{cuttings} = (v_m - v_s) * C_a * \frac{\pi}{4}(D^2 - d^2) \qquad (9-12)$$

In equilibrium,

$$\frac{ROP * \pi * D_b^2}{4} = (v_m - v_s) * C_a * \frac{\pi}{4}(D^2 - d^2) \qquad (9-13)$$

$$C_a = \frac{ROP * D_b^2}{(v_m - v_s)(D^2 - d^2)} \qquad (9-14)$$

From Equation (9-10),

$$T = \frac{v_m - v_s}{v_m}$$

$$v_m - v_s = T * v_m \qquad (9-15)$$

Substituting Equation (9-15) into (9-14),

$$C_a = \frac{ROP * D_b^2}{T * v_m (D^2 - d^2)} \qquad (9-16)$$

The velocity of the mud is given by

$$\bar{v} = \frac{Q}{A}$$

or

$$v_m = \frac{Q4}{\pi(D^2 - d^2)}$$

Thus

$$C_a = \frac{ROP * D_b^2}{T(D^2 - d^2)} * \frac{\pi(D^2 - d^2)}{4Q}$$

$$C_a = \frac{ROP * \pi * D_b^2}{4 * Q * T} \qquad (9\text{-}17)$$

A criterion commonly quoted for effective hole cleaning is a maximum cuttings concentration of .04 or .05 (4 or 5%). The greatest cuttings concentration will be obtained in the largest annular section, which is the riser or conductor pipe.

The cuttings often have a large effect upon the effective circulating density. For each section of the annulus, the total pressure is

$$\text{pressure} = \text{length} * \text{density} * g$$

The change in pressure due to the cuttings depends upon the density of the cuttings and the concentration of the cuttings in that section.

$$\Delta P_c = L_v * g * (\rho_{cuttings} - \rho_{mud}) * C_a \qquad (9\text{-}18)$$

The total pressure change added by the cuttings is therefore the sum of the pressure changes for each section.

$$\Sigma \Delta P = g * (\rho_{cuttings} - \rho_{mud}) * \Sigma(\Delta \text{length} * C_a) \qquad (9\text{-}19)$$

The effective circulating density then becomes

$$ECD_c = \text{hydrostatic pressure} + \Delta P_{friction} + \Delta P_{cuttings}$$

$$= (D_v - fl) * \rho * g + \Delta P_{friction} + \Delta P_{cuttings}$$

$$= MW + \frac{\Delta P_{friction} + \Delta P_{cuttings}}{(D_v - Fl) * g} \tag{9-20}$$

9.2 REFERENCES

1. Hannah, R. R., and L. J. Harrington, "Measurement of Dynamic Proppant Fall Rates in Fracturing Gels using a Concentric Cylinder Tester," J.P.T., May 1981.

2. Hopkin, E. A., "Factors Affecting Cuttings Removal During Rotary Drilling, J.P.T., June 1967.

3. Moore, P. L., Drilling Practices Manual, Petroleum Publishing Co., 1974.

4. Randall, B. V., and D. B. Anderson, "Flow of Mud During Drilling Operations," S.P.E. Paper 9444, 1980.

5. Sample, K. J., and A.T. Bourgoyne, "An Experimental Evaluation of Correlations Used for Predicting Cutting Slip Velocity," S.P.E. Paper 6645, 1977.

6. Sample, K. J., and A. T. Bourgoyne, "Develoment of Improved Laboratory and Field Procedures for Determining the Carrying Capacity of Drilling Fluids," S.P.E. Paper 7497, 1978.

7. Sifferman, T. R., G. M. Myers, E. L. Haden and H. A. Wahl, "Drill-Cuttings Transport in Full Scale Vertical Annuli," J.P.T., November 1974.

8. Walker, R. E., "Migration of Particles to a Hole Wall in a Drilling Well," S.P.E. J., June 1969.

9. Walker, R. E., and T. M. Mayes, "Design of Muds for Carrying Capacity," J.P.T., July 1975.

10. Williams, C. E., and G. H. Bruce, "Carrying Capacity of Drilling Muds," Trans. A.I.M.E. <u>192</u>, 1951.

11. Zeidler, H. U., "A Study of Fluid Hydraulics in Clear Water Drilling," Drilling Canada, September/October 1981.

10
OPTIMIZING THE HYDRAULICS PROGRAM

10.1 GENERAL

In an optimized drilling-hydraulics program the variables are controlled in such a manner that the well can be completed safely, with minimum damage to borehole formation and at the lowest possible cost. To achieve this, fluid density should just slightly exceed the hydrostatic pressure in the formation being drilled in permeable zones. Therefore, safety considerations usually dictate the minimum fluid density that can be used. A minimum fluid density also helps to control sloughing shales which otherwise could lead to expensive reaming or fishing jobs.

To minimize formation damage, it is necessary to reduce surges of pressure against the formation. Low gel strengths will aid in maintaining borehole stability, minimize the trip margin and permit drilling with fluid close to the minimum density, thereby improving penetration rate and reducing costs. However, the fluid must be able to suspend cuttings adequately, so viscous properties will be a compromise. Certain formations may require increased gel strength and viscosity in order to suspend cuttings and prevent pack-offs during connections. Wireline logging and testing programs will give better results if the hole condition is good and if invasion is not excessive.

After determining the desired fluid density and viscous properties, the engineer can then calculate the optimum flowrate and nozzle sizes. Minimum flowrate is usually dictated by annular velocity in the riser or surface casing. Eight meters per minute (25 ft/min) is often selected as a minimum, although the figure depends on penetration rate, hole size, and cuttings slip velocity. If drilling is fast and produces large cuttings, a minimum annular velocity higher than 8m/min might be indicated. On offshore installations, the annular velocity in long risers can be substantially increased from the BOPs by using booster pumps.

The upper limit on flowrate is often chosen to correspond to critical velocity opposite the drill collars. Keeping the annular flow laminar will usually minimize formation damage; however, in some hard-rock areas this limitation may not apply, and flowrate may then be limited only by the pump output.

Minimum and maximum flowrates will dictate the choice of pump liner size which, in turn, will impose an upper limit on the standpipe pressure. With the constraints of minimum and maximum flowrates and maximum standpipe pressure, exact flowrate and nozzle sizes can be selected in order to maximize a given parameter. It is usual to maximize either the hydraulic power at the bit or the jet impact force developed at the bit.

The parasitic pressure loss (the circulation-pressure loss in all parts of the system, except the bit) is an important quantifier in an optimized drilling hydraulics program. It can be calculated by totaling all pressure loss from the drillstring, the annulus, and that due to cuttings suspended in the annulus. Another method for calculating parasitic pressure loss is to subtract the bit pressure drop from the total standpipe pressure.

Hydraulic power is the product of pressure and flowrate. Figure 10-1 is a graph of hydraulic power versus flowrate: the straight line represents the power required at the surface to maintain the maximum standpipe pressure. For each flowrate, the power available at the bit is the hydraulic power at the surface, minus parasitic power loss:

$$W_b = P_s * Q - P_p * Q \qquad (10-1)$$

where

W_b = power available at bit
P_s = pressure at surface
Q = bulk flowrate
P_p = parasitic pressure loss

Maximum power at the bit occurs when the first derivative of this function with respect to flowrate is zero. The required flowrate, defined by

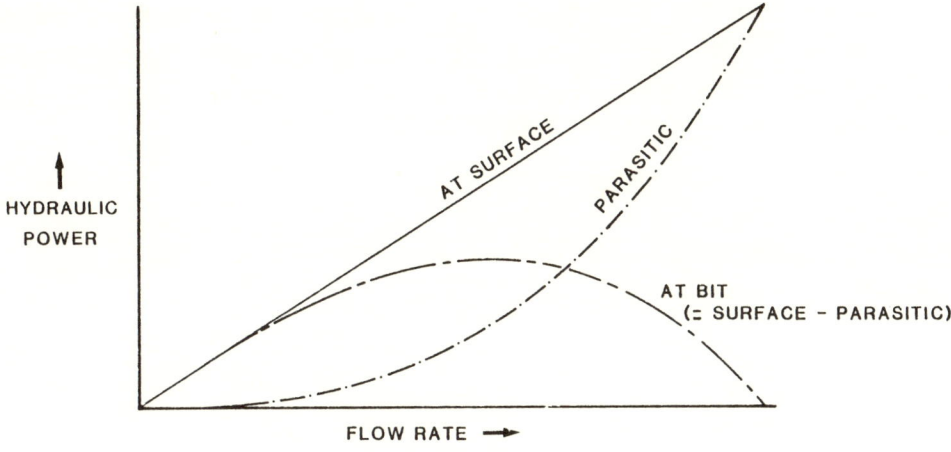

Figure 10-1. Hydraulic Power Vs. Flowrate

$$P_s - P_p - Q \frac{d}{dQ} P_p = 0 \qquad (10\text{-}2)$$

Equation (10-2) is a general criterion for finding the flowrate corresponding to maximum power at the bit. It is frequently simplified by assuming that the parasitic pressure loss is a power function of flowrate, with exponent m. The expression $Q \frac{d}{dQ} P_p$ then becomes $m * P_p$, and Equation (10-2) is simplified to

$$P_p = \frac{P_s}{(m+1)} \qquad (10\text{-}3)$$

where

P_p = parasitic pressure loss
P_s = pressure at surface
m = parasitic pressure loss exponent

The required flowrate, therefore, is that at which the parasitic pressure loss is a fraction [1/(m+1)] of the surface pressure. Knowing actual pressure loss at two flowrates gives sufficient information to calculate m and the optimum flowrate.

Another common assumption is to fix m at 1.86. Since the majority of the parasitic pressure loss occurs in the surface equipment and drillstring, where flow is turbulent, this approximation is seldom in serious error. It corresponds to 35 percent of total pressure being parasitic, or 65 percent of total surface pressure being developed at the bit.

A similar technique can be used to find optimum flowrate to maximize jet impact force. The impact force can be found by multiplying fluid density, flowrate and velocity. Thus,

$$F_i = \rho Q v_j \qquad (10\text{-}4)$$

where

F_i = impact force
ρ = fluid density
Q = bulk flowrate
v_j = jet velocity

Expressing jet velocity in terms of the bit pressure drop, $(P_s - P_p)$, this becomes

$$F_i = \rho * Q * C_v \left[\frac{2(P_s - P_p)}{\rho} \right]^{.5} \qquad (10\text{-}5)$$

where

F_i = impact force
ρ = fluid density
Q = bulk flowrate
C_v = velocity coefficient
P_s = pressure at surface
P_p = parasitic pressure loss

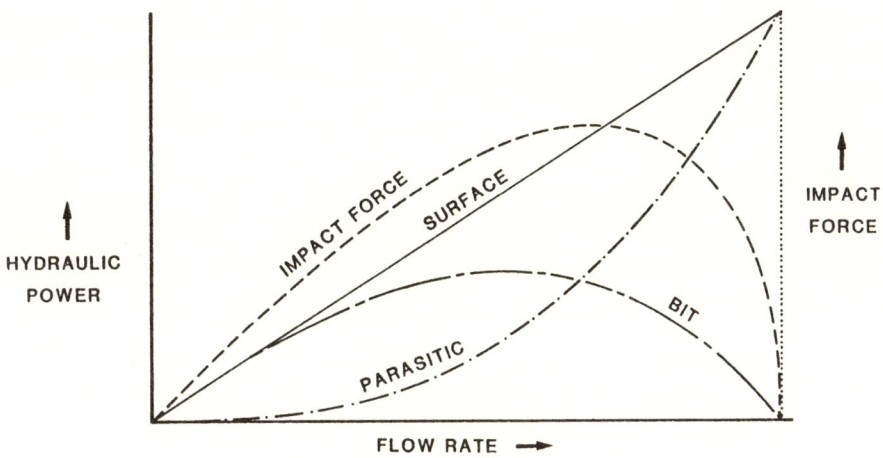

Figure 10-2. Optimum Impact Force

From Figure 10-2 it can be seen that, when the first derivative of impact force with respect to flowrate is set to zero, the optimum flowrate can be found. Differentiating Equation (10-5) with respect to Q, equating to zero, and rearranging,

$$2(P_s - P_p) - Q \frac{d}{dQ} P_p = 0 \qquad (10-6)$$

This is the condition to be fulfilled by the flowrate for maximum impact force. Again, assuming that P_p is a power function of Q,

$$P_p = P_s * \frac{2}{(m + 2)} \qquad (10-7)$$

If m is assumed to be 1.86, the optimum flowrate develops 52 percent of the surface pressure as a parasitic loss, leaving 48 percent across the bit nozzles.

At this stage, the flowrate giving maximum hydraulic power or impact force has been found. If this flowrate lies outside the maximum and minimum limits, the appropriate limiting value is used, and parasitic pressure loss at the limiting value is subtracted from surface pressure to give the desired pressure drop at the bit.

It is possible to design a computer program to ensure that hydraulic power does not exceed a specified limit. Some operators prefer to limit power to five horsepower per square inch of hole area, or a similar value. This limitation requires a trial-and-error solution if calculations are made manually.

It remains only to select the nozzle sizes which will give the desired pressure drop at the bit at the required flowrate. The pressure drop can be converted to equivalent jet velocity by use of Equation (6-5), and Equation (6-2) then gives the total nozzle area. This area may be obtained in various ways since jet nozzles need not be all of the same size.

It is sometimes necessary to run a bit with large nozzles in order to circulate lost-circulation material. This frequently causes a great decrease in hydraulic power and impact force. However, in many cases it is possible to maintain reasonably good hydraulics by reducing the number of nozzles. For example, two 16/32" nozzles have approximately the same area as three 13/32" nozzles. Reducing the number of nozzles from three to two boosts hydraulic power by a factor of 2.25, and increases impact force by 50 percent.

10.2 REFERENCES

1. Allen, J. H., and W. Baker, "Improved Drilling Performance with Enhanced Crossflow Rock Bits," J.P.T., October 1980.

2. Arbizu, A. G., and R. McGinnis, "Computer Output Compares Hydraulic Maximizing Estimates with Laminar Flow Constraints," Pet. Engr., November 1980.

3. Bourgoyne, A. T., and F. S. Young, "A Multiple Regression Approach to Optimal Drilling and Abnormal Pressure Detection," S.P.E. J., August 1974.

4. Buckley, P., and R. A. Jardiolin. "How to Simplilfy Rig Hydraulics," Pet. Engr., March 1982.

5. Clark, R. K., and J. E. Fontenot, "Field Measurements of the Effects of Drillstring Velocity, Pump Speed and Lost Circulation Material on Downhole Pressures," S.P.E. Paper 4970, 1974.

6. Galloway, L. A., C. B. Lowrey and R. J. Link, "Energy Conservaton and Polymer Drilling Fluids," Pet. Engr., August 1980.

7. Kendall, H. A., and W. C. Goins, "Design and Optimization of Jet-Bit Programs for Maximum Hydraulic Horsepower, Impact Force or Jet Velocity." Pet. Trans. A.I.M.E. 219, 1960.

8. McLean, R. H., "Crossflow and Impact Under Jet Bits," J.P.T., November 1964.

9. Miska, S., and P. Skalle, "Theoretical Description of a New Method of Optimal Program Design," S.P.E. J., August 1981.

10. Robinson, L., "On-Site Nozzle Selection Increases Drilling Performance," Pet. Engr., December 1981.

11. Siegel, B., "Hand-Held Calculators Augment Optimized Drilling Programs," Pet. Engr., October 1980.

12. Smalling, D. A., and T. A. Key, "Optmization of Jet-Bit Hydraulics Using Impact Pressure," S.P.E.Paper 8840, 1979.

13. Sutko, A. A., and G. M. Meyers, "The Effect of Nozzle Size, Number and Extension on the Pressure Distribution Under a Tricone Bit," J.P.T., November 1971.

14. Zeidler, H. U., "A Study of Fluid Hydraulics in Clear Water Drilling," Drilling Canada, September/October 1981.

APPENDIX A
NOMENCLATURE

A .... Area

A_n .... Area, total nozzle

α .... Annulus-diameter ratio

B .... Bingham number

b .... Ratio, radius

C .... Number of pump cylinders

C_a ...... Cuttings concentration in annulus

C_D .... Drag coefficient

C_I .... Inlet coefficient

C_o .... Outlet coefficient

D .... Diameter, outer

d .... Diameter, inner

D_b .... Diameter, bit

d_I .... Diameter, inside drillstring

d_p .... Diameter, particle

D_v .... Depth, vertical

Σ .... Summation symbol

ϵ .... Roughness, absolute

ξ .... Pump efficiency

ECD .... Density, Equivalent Circulating

ECD_c .... Density, Equivalent Circulating with cuttings effect

F .... Force

F_I .... Force, hydraulic impact

Fl ...... Distance (depth), RKB to flowline

f .... Friction factor

G .... Geometric factor

G_v .... Viscometer correction factor

g .... Gravitational acceleration

γ .... Shear rate

γ_β .... Shear rate at bob

γ_o .... Shear rate intercept

He .... Hedstrom number

k .... Consistency factor

K_s ...... Coefficient, surface pressure loss

K_c .... Constant, clinging

K_t ..... Constant, turbulent flow, clinging

L .... Length

L_v .... Length, vertical

λ .... Dimensionless radius

M .... Moment

m .... Parasitic pressure loss exponent

μ .... Viscosity

μ_∞ .... Plastic viscosity

$\bar{\mu}$ .... Mean viscosity

θ .... Viscometer dial reading

MW .... Mudweight
N .... Rotation speed
N_p .... Pump-stroke rate
n .... Flow behavior index
σ .... Cuttings concentrations
P .... Pressure
ρ .... Density
$ρ_f$ .... Fluid Density
$ρ_p$ .... Particle Density
P_a .... Pressure loss, annular
P_b .... Pressure drop across bit
P_g .... Gel-breaking pressure
P_h .... Pressure, hydrostatic
P_p .... Pressure loss, parasitic
P_s .... Pressure, circulating at surface
ΔP .... Pressure differential
$ΔP_b$ .... Pressure loss, total across bit
$ΔP_c$ .... Pressure change due to cuttings
$ΔP_I$ ..... Pressure loss at restriction inlet
$ΔP_J$ ..... Pressure loss total at tool joint
$ΔP_L$ .... Pressure loss, laminar flow
$ΔP_o$ .. Pressure loss at restriction outlet
$ΔP_{or}$ .... Pressure loss through an orifice
$ΔP_t$ .... Pressure loss at orifice due to turbulence
$ΔP_T$ .... Pressure loss, turbulent flow
PV .... Plastic viscosity
Q .... Flowrate, bulk

ρ .... Density
λ .... Radius, dimensionless
R .... Radius, outer
r .... Radius, inner
$r_β$ .... Radius of the bob
Re .... Reynolds number
Re_c .... Reynolds number, critical
Re_e .... Reynolds number equivalent
Re_L .... Reynolds number, laminar flow
Re_p .... Reynolds number particle
Re_T .... Reynolds number, turbulent flow
ROP .... Rate of penetration
σ .... Cuttings concentration
T .... Ratio, cuttings transport
$τ_g$ .... Gel strength
τ .... Shear stress
$τ_β$ .... Shear stress at bob
$τ_o$ .... Yield stress
$τ_w$ .... Shear stress at wall
V .... Volume
v .... Velocity
$\bar{v}$ .... Velocity, average
v_c .... Velocity, critical
v_{eff} .... Velocity, effective
v_J .... Velocitoy, jet
v_L .... Velocity, cuttings, net
v_m .... Velocity, annular mud

v_p .... Velocity, pipe

v_s .... Velocity, particle slip

W_b .... Power at the bit

w ... Width of parallel-plate flow channel

ω .... Angular velocity

x_c .... Dimensionless reciprocal pressure at the critical point

YP............................ Yield point

y,z .... Dummy variables

Z .... Stability parameter

Absolute roughness	ε	Diameter (annulus/diam. ratio)	α
Acceleration due to gravity	g	Diameter, bit	D_b
Angular velocity	ω	Diameter, inner	d
Annular mud velocity	v_m	Diameter, inside drillstring	d_i
Annular pressure loss	P_a	Diameter, outer	D
Annulus/diameter ratio	α	Diameter, particle	d_p
Area	A	Differential pressure	ΔP
Area, total nozzle	A_n	Dimensionless radius	λ
Average velocity	$\bar{v}$	Dimensionless reciprocal pressure drop at the critical point	x_c
Bingham number	B	Distance from RKB to flowline	Fl
Bit diameter	D_b	Drag coefficient	C_D
Bit (total press. loss across)	ΔP_b	Dummy variables	y,z
Bulk flowrate	Q	Effective velocity	v_{eff}
Circulating press. at surface	P_s	Efficiency, pump	ξ
Coefficient, outlet	C_o	Equivalent Circulating Density	ECD
Consistency factor	k	Equivalent Reynolds number	Re_e
Constant, clinging	K_c	Flow behavior index	n
Constant, turbulent flow, clinging	K_t	Fluid density	ρ_f
Cuttings concentration	σ	Flowline (distance from RKB)	Fl
Cuttings concentration in annul.	C_a	Flowrate	Q
Cuttings transport ratio	T	Force	F
Density	ρ	Friction factor	f
Density, Equivalent Circulating	ECD	Gel-breaking pressure	P_g
Depth, vertical	D_v	Gel strength	τ_g
Depth, RKB to flowline	Fl	Geometric factor	G
Dial reading	θ		

Gravitational acceleration	g	Particle density	ρ_p
Hedstrom number	He	Particle diameter	d_p
Hydraulic impact force	F_I	Particle Reynolds number	Re_p
Hydrostatic Pressure	P_h	Particle slip velocity	v_s
Inlet coefficient	C_i	Particle thickness	h
Inner radius	r	Penetration, rate of	ROP
Inside diam. of drillstring	d_i	Pipe velocity	v_p
Jet velocity	v_j	Plastic viscosity	μ_∞
Laminar pressure loss	ΔP_L	Plastic viscosity	ρ_V
Length	L	Power at the bit	W_b
Length, vertical	L_v	Pressure	P
Mean viscosity	$\bar{\mu}$	Pressure loss at restriction inlet	ΔP_i
Moment	M	Pressure loss at restriction outlet	ΔP_o
Mudweight	MW	Pressure change due to cuttings	ΔP_c
Number, Bingham	B	Pressure, circulating, at surface	P_s
Number, Hedstrom	He	Pressure differential	ΔP
Number, Reynolds	Re	Pressure, hydrostatic	P_h
Number, Reynolds (equivalent)	Re_e	Pressure loss, annular	P_a
Number, Reynolds (particle)	Re_p	Pressure loss total at tool joint	ΔP_j
Number of pump cylinders	C	Pressure loss, laminar flow	ΔP_L
Outer diameter	D	Pressure loss thru an orifice	ΔP_{or}
Outer radius	R	Pressure loss, parasitic	P_p
Outlet coefficient	C_o	Pressure loss at orifice due to turbulence	ΔP_t
Parameter, stability	Z	Pressure loss, turbulent flow	ΔP_T
Parasitic press. loss	P_p	Pressure loss parasitic exponent	m
Parasitic press. loss exponent	m	Pressure loss total across the bit	ΔP_b

Pump cylinders number	C
Pump efficiency	ξ
Pump-stroke rate	N_p
Radius, dimensionless	λ
Radius, inner	r
Radius, outer	R
Radius of bob	r_β
Rate of Penetration	ROP
Ratio, annulus diameter	α
Ratio, cuttings transport	T
Ratio, radius	b
Reynolds number	Re
Reynolds number, critical	Re_c
Reynolds number equivalent	Re_e
Reynolds number, laminar flow	Re_L
Reynolds number, particle	Re_p
Reynolds number, turbulent flow	Re_T
Rotation speed	N
Roughness, absolute	ε
Shear rate	γ
Shear rate at bob	γ_β
Shear rate intercept	γ_o
Shear stress	τ
Shear stress at bob	τ_β
Shear stress at wall	τ_w
Slip velocity	v_s
Stability parameter	Z
Stress, shear	τ
Stress, yield	τ_o
Summation symbol	Σ
Thickness	h
Total press. loss across the bit	ΔP_b
Turbulent flow clinging constant	K_t
Velocity	v
Velocity, angular	ω
Velocity, annular mud	v_m
Velocity, average	$\bar{v}$
Velocity, effective	v_{eff}
Velocity, jet	v_j
Velocity, critical	v_c
Velocity, cuttings, net	v_L
Velocity, pipe	v_p
Viscometer correction factor	G_v
Viscometer dial reading	θ
Viscosity	μ
Viscosity, plastic	μ_∞
Volume	V
Width of parallel-plate flow channel	w
Yield point	YP
Yield stress	τ_o

APPENDIX B
UNITS

Equations developed in the text assume that all quantities are measured in a consistent set of units. Key equations are presented in Appendix C, for both S.I. units and common oilfield units. The units of measurement are as follows:

Quantity	S.I. Unit	Oilfield Unit
A, Area	meter^2	inch^2
D, Diameter	meter	inch
D_v, Vertical Depth	meter	ft
d, Diameter	meter	inch
ECD, Equivalent Circ. Density	kilogram-meter^{-3}	lb/gal
g, Acceleration	9.807 meter-sec^{-2}	32.17 ft/sec^2
h, Thickness	meter	inch
k, Consistency Factor	pascal-sec^n	lb-sec^n/100 ft^2
L, Length	meter	ft
M, Moment	newton-meter	lb-inch
N, Rotation Speed	RPM	RPM
Np, Pump Stroke Rate	strokes/sec	strokes/min
P, Pressure	pascal	psi
Q, Flowrate	meter^3-sec^{-1}	gal/min
R, Radius	meter	inch
ROP, Penetration Rate	meter-sec^{-1}	ft/hr
r, Radius	meter	inch

Quantity	S.I. Unit	Oilfield Unit
S, Spring Factor	newton-meter-degree^{-1}	lb-inch/degree
V, Volume	meter^3	gal
v, Velocity	meter-sec^{-1}	ft/min
w, Width	meter	inch
γ, Shear Rate	second^{-1}	sec^{-1}
θ, Dial Reading	degree	degree
μ, Viscosity	pascal-second	centipoise
ρ, Density	kilogram-meter^{-3}	lb/gallon
τ, Shear Stress	pascal	lb/100 foot^2
ω, Angular Velocity	degree-sec^{-1}	degree/sec
W_b, Power at Bit	watt	horsepower
F, Force	newton	lb

APPENDIX C
KEY EQUATIONS

S.I. Units

Pump Output	C-1
Rheometry, Bingham	C-1
Rheometry, Power Law	C-1
Pipe Flow, Bingham (exact)	C-2
Pipe Flow, Bingham (approximate)	C-3
Pipe Flow, Power Law	C-4
Parallel-Plate Flow, Bingham (exact)	C-5
Parallel-Plate Flow, Bingham (approximate)	C-6
Parallel-Plate Flow, Power Law	C-7
Annular Flow, Power Law	C-9
Tool Joints	C-10
Bit Nozzles	C-11
Cuttings Transport	C-11
Hydraulic Power	C-11
Impact Force	C-11

Oilfield Units

Pump Output	C-12
Rheometry, Binhgam	C-12
Rheometry, Power Law	C-12
Pipe Flow, Bingham (exact)	C-13
Pipe Flow, Bingham (approximate)	C-14
Pipe Flow, Power Law	C-15
Parallel Plate Flow, Bingham (exact)	C-16
Parallel-Plate Flow, Bingham (approximate)	C-17
Parallel-Plate Flow, Power Law	C-18
Annular Flow, Power Law	C-19
Tool Joints	C-20
Bit Nozzles	C-20
Cuttings Transport	C-21
Hydraulic Power	C-21
Impact Force	C-21

The commonly required equations are grouped here for easy reference. Units of measurement are given in Appendix B. When the "oilfield" version of an equation is omitted, it is identical to the SI version.

Pump Output

$$V = \frac{\pi}{4} LC \xi (2D^2 - d^2) \tag{4-2}$$

$$V = \frac{\pi}{4} LC \xi D^2 \tag{4-1}$$

$$Q = VN_p \tag{4-3}$$

Rheometry, Bingham

$$\mu_\infty = \frac{(\theta_2 - \theta_1)}{1000} \tag{2-37}$$

$$\tau_o = .4788 (2\theta_1 - \theta_2) \tag{2-38}$$

Rheometry, Power Law

$$n = \frac{\log(\theta_2/\theta_1)}{\log 2} \tag{2-29}$$

$$k = .511 \, \theta_1 \left[\frac{n}{20\pi} (1 - 1.0678^{-2/n}) \right]^n \tag{2-39}$$

Pipe Flow, Bingham (exact)

$$\bar{v} = \frac{4Q}{\pi D^2} \qquad (4-4)$$

$$\beta = 1 + \frac{6\mu_\infty \bar{v}}{\tau_o D} \qquad (5-21)$$

$$z = \left(\beta^2 + \sqrt{\beta^4 - 1}\right)^{1/3} \qquad (5-22)$$

$$y = 2\left(z + \frac{1}{z}\right) \qquad (5-23)$$

$$x = \frac{1}{2}\left(\sqrt{y} - \sqrt{\frac{8\beta}{\sqrt{y}} - y}\right) \qquad (5-24)$$

$$\Delta P_L = \frac{4L\tau_o}{xD} \qquad (5-25)$$

$$Re_c = 2000$$

$$\bar{\mu} = \frac{\Delta P_L\, D^2}{32\, L\bar{v}} \qquad (5-28)$$

$$Re = \frac{\rho \bar{v} D}{\bar{\mu}} \qquad (5-27)$$

$$\Delta P = \Delta P_L \qquad (Re \leqslant Re_c)$$

$$f = \frac{.079}{Re^{.25}} \qquad (Re > Re_c) \qquad (5-36)$$

$$\Delta P = \Delta P_T = \frac{2fL\rho \bar{v}^2}{D} \qquad (Re > Re_c) \qquad (5-32)$$

<u>Pipe Flow, Bingham (approximate)</u>

$$\bar{v} = \frac{4Q}{\pi D^2} \qquad (4-4)$$

$$\Delta P_L \simeq \frac{32L\mu_\infty \bar{v}}{D^2} + \frac{16L\tau_o}{3D} \qquad (5-26)$$

$$Re_c = 2000$$

$$\bar{\mu} = \frac{\Delta P_L D^2}{32L\bar{v}} \qquad (5-28)$$

$$Re = \frac{\rho \bar{v} D}{\bar{\mu}} \qquad (5-27)$$

$$\Delta P = \Delta P_L \qquad (Re < Re_c)$$

$$f = \frac{.079}{Re^{.25}} \qquad (Re > Re_c) \qquad (5\text{-}36)$$

$$\Delta P = \Delta P_T = \frac{2fL\rho\bar{v}^2}{D} \qquad (Re > Re_c) \qquad (5\text{-}32)$$

$$v_c = \frac{1000\mu_\infty + 1000\sqrt{\mu_\infty^2 + \frac{\rho\tau_0 D^2}{3000}}}{\rho D}$$

Pipe Flow, Power Law

$$\bar{v} = \frac{4Q}{\pi D^2} \qquad (4\text{-}4)$$

$$\Delta P_L = \frac{4Lk}{D}\left[\frac{8\bar{v}}{D} * \frac{(3n+1)}{4n}\right]^n \qquad (5\text{-}18)$$

$$Re_c = 3470 - 1370n$$

$$\bar{\mu} = \frac{\Delta P_L D^2}{32L\bar{v}} \qquad (5\text{-}28)$$

$$Re = \frac{\rho \bar{v} D}{\bar{\mu}} \qquad (5\text{-}27)$$

$$\Delta P = \Delta P_L \quad (Re < Re_c)$$

$$y = (\log n + 3.93)/50 \tag{5-37}$$

$$z = (1.75 - \log n)/7 \tag{5-38}$$

$$f = y * Re^{-z} \quad (Re \geq Re_c + 800) \tag{5-36}$$

$$f = \frac{16}{Re} + \frac{(Re - Re_c)}{800}(y * Re^{-z} - \frac{16}{Re}) \quad (Re_c < Re < \lfloor Re_c + 800 \rfloor) $$

$$\Delta P = \Delta P_T = \frac{2fL\rho \bar{v}^2}{D} \quad (Re > Re_c) \tag{5-32}$$

$$v_c = \left(\frac{k\, Re_c}{8\rho}\right)^{\frac{1}{(2-n)}} * \left[\frac{8(3n+1)}{D * 4n}\right]^{\frac{n}{(2-n)}}$$

Parallel-Plate Flow, Bingham (exact)

$$\bar{v} = \frac{4Q}{\pi D^2 (1 - \alpha^2)} \tag{4-4}$$

$$B = \frac{\tau_o D(1 - \alpha)}{\mu_\infty \bar{v}} \tag{7-21}$$

$$\beta = \left(\frac{B}{B + 8}\right)^{.5} \tag{7-22}$$

$$x = \frac{2}{\beta} \sin\left(\frac{1}{3}\sin^{-1}(\beta^3)\right) \tag{7-23}$$

$$\Delta P_L = \frac{4L\tau_0}{x * D(1-\alpha)} \tag{7-24}$$

$$Re_c = 3000$$

$$\bar{\mu} = \frac{\Delta P_L D^2 (1-\alpha)^2}{48L\,\bar{v}} \tag{7-32}$$

$$Re = \frac{\rho\,\bar{v}\,D\,(1-\alpha)}{\bar{\mu}} \tag{7-30}$$

$$\Delta P = \Delta P_L \quad (Re < Re_c)$$

$$f = \frac{.079}{Re^{.25}} \quad (Re > Re_c) \tag{5-36}$$

$$\Delta P = \Delta P_T = \frac{2fL\rho\bar{v}^2}{D(1-\alpha)} \quad (Re > Re_c) \tag{5-32}$$

Parallel-Plate Flow, Bingham (approximate):

$$\bar{v} = \frac{4Q}{\pi D^2 (1-\alpha^2)} \tag{4-4}$$

$$\Delta P_L \simeq \frac{48L\,\mu_\infty\,\bar{v}}{D^2(1-\alpha)^2} + \frac{6L\,\tau_0}{D(1-\alpha)} \tag{7-25}$$

$$Re_c = 3000$$

$$\bar{\mu} = \frac{\Delta P_L D^2 (1-\alpha)^2}{48\,L\,\bar{v}} \tag{7-32}$$

$$Re = \frac{\rho \bar{v} D (1 - \alpha)}{\mu} \qquad (7\text{-}30)$$

$$\Delta P = \Delta P_L \qquad (Re \leq Re_c)$$

$$f = \frac{.079}{Re^{.25}} \qquad (Re > Re_c) \qquad (5\text{-}36)$$

$$\Delta P = \Delta P_T = \frac{2fL \rho \bar{v}^2}{D(1 - \alpha)} \qquad (Re > Re_c) \qquad (5\text{-}32)$$

$$v_c = \frac{1500 \mu_\infty + 1500 \sqrt{\mu_\infty + \frac{\rho \tau_0 D^2 (1-\alpha)^2}{6000}}}{\rho D(1 - \alpha)}$$

Parallel-Plate Flow, Power Law

$$\bar{v} = \frac{4Q}{\pi D^2 (1 - \alpha^2)} \qquad (4\text{-}4)$$

$$\Delta P_L = \frac{4L k}{D(1 - \alpha)} \left[\frac{12 \bar{v} (2n + 1)}{D(1 - \alpha) \, 3n} \right]^n \qquad (7\text{-}19)$$

$$Re_c = \frac{3}{2} (3470 - 1370 n)$$

$$\bar{\mu} = \frac{\Delta P_L \, D^2 (1 - \alpha)^2}{48 \, L \, \bar{v}} \qquad (7-32)$$

$$Re = \frac{\rho \, \bar{v} \, D(1 - \alpha)}{\bar{\mu}} \qquad (7-30)$$

$$\Delta P = \Delta P_L \qquad (Re \leq Re_c)$$

$$y = \frac{(\log n + 3.93)}{50} \qquad (5-37)$$

$$z = \frac{(1.75 - \log n)}{7} \qquad (7-38)$$

$$f = y * Re^{-z} \qquad (Re > Re_c + 1200) \qquad (5-36)$$

$$f = \frac{24}{Re} + \frac{(Re - Re_c)}{1200} (y * Re^{-z} - \frac{24}{Re}) \quad (Re_c < Re < [Re_c + 1200])$$

$$\Delta P = \Delta P_T = \frac{2 f L \rho \, \bar{v}^2}{D(1 - a)} \qquad (Re > Re_c) \qquad (5-32)$$

$$v_c = \left(\frac{k * Re_c}{12\rho}\right)^{\frac{1}{(2-n)}} \left[\frac{12(2n + 1)}{D(1 - \alpha) \, 3n}\right]^{\frac{n}{(2-n)}}$$

Annular Flow, Power Law

$$\bar{v} = \frac{4Q}{\pi D^2 (1 - \alpha^2)} \qquad (4\text{-}4)$$

$$y = .37 \, n^{-.14} \qquad (7\text{-}26)$$

$$z = 1 - (1 - \alpha^y)^{1/y} \qquad (7\text{-}27)$$

$$G = \left(1 + \frac{z}{2}\right) \left[\frac{(3 - z)n + 1}{(4 - z)n}\right] \qquad (7\text{-}28)$$

$$\Delta P_L = \frac{4 \, k \, L}{D(1 - \alpha)} \left[\frac{8 \, \bar{v} \, G}{D(1 - \alpha)}\right]^n \qquad (7\text{-}29)$$

$$G_N = \frac{1 + \alpha^2 + \frac{(1 - \alpha^2)}{\ln \alpha}}{(1 - \alpha)^2}$$

$$\bar{\mu} = \frac{\Delta P_L \, D^2 (1 - \alpha)^2 \, G_N}{32 \, L \, \bar{v}} \qquad (7\text{-}31)$$

$$Re = \frac{\rho \, \bar{v} \, D(1 - \alpha)}{\bar{\mu}} \qquad (7\text{-}30)$$

$$Re_c = \frac{(3470 - 1370 \, n)}{G_N}$$

$$\Delta P = \Delta P_L \qquad (Re \leq Re_c)$$

$$y = (\log n + 3.93) / 50 \tag{5-37}$$

$$z = (1.75 - \log n) / 7 \tag{5-38}$$

$$f = y * Re^{-z} \quad (Re > Re_c + \frac{800}{G_N}) \tag{5-36}$$

$$f = \frac{16}{Re * G_N} + (Re - Re_c) \frac{G_N}{800} \left(y * Re^{-z} - \frac{16}{Re * G} \right) \quad (Re_c < Re < \lfloor Re_c + \frac{800}{G_N} \rfloor)$$

$$\Delta P = \Delta P_T = \frac{2fL\rho \bar{v}^2}{D(1-\alpha)} \quad (Re > Re_c) \tag{5-32}$$

$$v_c = \left(\frac{k * Re_c * G_N}{8\rho} \right)^{\frac{1}{(2-n)}} * \left[\frac{8G}{D(1-\alpha)} \right]^{\frac{n}{(2-n)}}$$

Tool Joints

$$\Delta P_j \simeq \frac{\rho}{2} (v_2^2 - v_1^2) \tag{5-46}$$

Bit Nozzles

$$v_j = \frac{4Q}{\pi \Sigma d_j^2} \quad (6\text{-}2)$$

$$\Delta P_b = \frac{\rho v_j^2}{1.805} \quad (6\text{-}5)$$

Cuttings Transport

$$Re_p = \frac{\rho_f v_s d_p}{\mu} \quad (9\text{-}1)$$

$$C_D = \frac{40}{Re_p} + 1 \quad (9\text{-}7)$$

$$v_s = \left[\frac{2hg(\rho_p - \rho_f)}{C_D \rho_f}\right]^{.5} \quad (9\text{-}3)$$

$$C_a = \frac{ROP * D_b^2}{(v_m - v_s) D^2 (1 - \alpha^2)} \quad (9\text{-}14)$$

Hydraulic Power

$$W_b = \Delta P_b * Q \quad (10\text{-}1)$$

Impact Force

$$F_i = \rho * Q * v_j \quad (10\text{-}5)$$

Pump Output

$$V = .0034 \, LC \, \xi \, (2D^2 - d^2) \qquad (4-2)$$

$$V = .0034 \, L \, C \, \xi \, D^2 \qquad (4-1)$$

$$Q = VN_p \qquad (4-3)$$

Rheometry, Bingham

$$\mu_\infty = \theta_2 - \theta_1 \qquad (2-37)$$

$$\tau_o = 2\theta_1 - \theta_2 \qquad (2-38)$$

Rheometry, Power Law

$$n = \frac{\log(\theta_2/\theta_1)}{\log 2} \qquad (2-29)$$

$$k = 1.067 \, \theta_1 \left[\frac{n}{20\pi} (1 - 1.0678^{-2/n}) \right]^n \qquad (2-39)$$

Pipe Flow, Bingham (exact)

$$\bar{v} = \frac{24.51 \, Q}{D^2} \tag{4-4}$$

$$\beta = 1 + \frac{\mu_\infty \bar{v}}{399 \, \tau_o D} \tag{5-21}$$

$$z = \left(\beta^2 + \sqrt{\beta^4 - 1}\right)^{1/3} \tag{5-22}$$

$$y = 2\left(z + \frac{1}{z}\right) \tag{5-23}$$

$$x = \frac{1}{2}\left(\sqrt{y} - \sqrt{\frac{8\beta}{\sqrt{y}} - y}\right) \tag{5-24}$$

$$\Delta P_L = \frac{L \, \tau_o}{300 * x * D} \tag{5-25}$$

$$Re_c = 2000$$

$$\bar{\mu} = \frac{90000 \, \Delta P_L D^2}{L \, \bar{v}} \tag{5-28}$$

$$Re = \frac{15.47 \, \rho \, \bar{v} \, D}{\bar{\mu}} \tag{5-27}$$

$$\Delta P = \Delta P_L \qquad (Re < Re_c)$$

$$f = \frac{.079}{Re^{.25}} \qquad (Re > Re_c) \tag{5-36}$$

$$\Delta P = \Delta P_T = \frac{f l \rho \, \bar{v}^2}{92903 \, D} \qquad (Re > Re_c) \tag{5-32}$$

Pipe Flow, Bingham, Approximate

$$\bar{v} = \frac{24.51 \, Q}{D^2} \tag{4-4}$$

$$\Delta P_L \simeq \frac{L \mu_\infty \bar{v}}{90000 \, D^2} + \frac{L \tau_0}{225 \, D} \tag{5-26}$$

$$Re_c = 2000$$

$$\bar{\mu} = \frac{90000 \, \Delta P_L \, D^2}{L \, \bar{v}} \tag{5-28}$$

$$Re = \frac{15.47 \, \rho \, \bar{v} \, D}{\bar{\mu}} \tag{5-27}$$

$$\Delta P = \Delta P_L \quad (Re \leq Re_c)$$

$$f = \frac{.079}{Re^{.25}} \quad (Re > Re_c) \tag{5-36}$$

$$\Delta P = \Delta P_T = \frac{f L \rho \bar{v}^2}{92903 \, D} \quad (Re > Re_c) \tag{5-32}$$

$$v_c = \frac{64.64 \, \mu_\infty + 64.64 \sqrt{\mu_\infty^2 + 12.34 \, \rho \, \tau_0 \, D^2}}{\rho \, D}$$

Pipe Flow, Power Law

$$\bar{v} = \frac{24.51 \, Q}{D^2} \qquad (4\text{-}4)$$

$$\Delta P_L = \frac{L \, k}{300 \, D} \left[\frac{1.6 \, \bar{v}}{D} * \frac{(3n+1)}{4n} \right]^n \qquad (5\text{-}18)$$

$$Re_c = 3470 - 1370n$$

$$\bar{\mu} = \frac{90000 \, \Delta P_L D^2}{L \, \bar{v}} \qquad (5\text{-}28)$$

$$Re = \frac{15.47 \, \rho \, \bar{v} \, D}{\bar{\mu}} \qquad (5\text{-}27)$$

$$\Delta P = \Delta P_L \qquad (Re \leq Re_c)$$

$$y = (\log n + 3.93) / 50 \qquad (5\text{-}37)$$

$$z = (1.75 - \log n) / 7 \qquad (5\text{-}38)$$

$$f = y * Re^{-z} \qquad (Re \geq Re_c + 800) \qquad (5\text{-}36)$$

$$f = \frac{16}{Re} + \frac{(Re - Re_c)}{800} \left(y * R_e^{-z} - \frac{16}{Re} \right) \qquad (Re_c < Re < [Re_c + 800])$$

$$\Delta P = \Delta P_T = \frac{f L \rho \, \bar{v}^2}{92903 \, D} \qquad (Re > Re_c) \qquad (5\text{-}32)$$

$$v_c = \left(\frac{19.33 \, k * Re_c}{\rho} \right)^{\frac{1}{(2-n)}} * \left[\frac{1.6(3n+1)}{D * 4n} \right]^{\frac{n}{(2-n)}}$$

Parallel-Plate Flow, Bingham (exact)

$$\bar{v} = \frac{24.51 \, Q}{D^2 (1 - \alpha^2)} \qquad (4\text{-}4)$$

$$B = \frac{2394 \, \tau_o \, D(1 - \alpha)}{\mu_\infty \, \bar{v}} \qquad (7\text{-}21)$$

$$\beta = \left(\frac{B}{B + 8}\right)^{.5} \qquad (7\text{-}22)$$

$$x = \frac{2}{\beta} \sin\left[\frac{1}{3} \sin^{-1}(\beta^3)\right] \qquad (7\text{-}23)$$

$$\Delta P_L = \frac{L \, \tau_o}{300 * x * D(1 - \alpha)} \qquad (7\text{-}24)$$

$$Re_c = 3000$$

$$\bar{\mu} = \frac{60000 \, \Delta P_L \, D^2 (1 - \alpha)^2}{L \, \bar{v}} \qquad (7\text{-}32)$$

$$Re = \frac{15.47 \, \rho \, \bar{v} \, D(1 - \alpha)}{\bar{\mu}} \qquad (7\text{-}30)$$

$$\Delta P = \Delta P_L \qquad (Re \leq Re_c)$$

$$f = \frac{.079}{Re^{.25}} \qquad (Re > Re_c) \qquad (5\text{-}36)$$

$$\Delta P = \Delta P_T = \frac{f \, L \, \rho \, \bar{v}^2}{92903 \, (D1 - \alpha)} \qquad (Re > Re_c) \qquad (5\text{-}32)$$

Parallel-Plate Flow, Bingham (approximate)

$$\bar{v} = \frac{24.51 \, Q}{D^2(1 - \alpha^2)} \tag{4-4}$$

$$\Delta P_L \approx \frac{L \mu_\infty \bar{v}}{60000 \, D^2 (1-\alpha)^2} + \frac{L \tau_0}{200 \, D(1-\alpha)} \tag{7-25}$$

$$Re_c = 3000$$

$$\bar{\mu} = \frac{60000 \, \Delta P_L D^2 (1 - \alpha)^2}{L \, \bar{v}} \tag{7-32}$$

$$Re = \frac{15.47 \, \rho \, \bar{v} \, D(1 - \alpha)}{\bar{\mu}} \tag{7-30}$$

$$\Delta P = \Delta P_L \qquad (Re \leq Re_c)$$

$$f = \frac{.079}{Re^{.25}} \qquad (Re > Re_c) \tag{5-36}$$

$$\Delta P = \Delta P_T = \frac{f \, L \, \rho \, \bar{v}^2}{92903 \, D(1-\alpha)} \qquad (Re > Re_c) \tag{5-32}$$

$$v_c = \frac{96.88 \mu_\infty + 96.88 \sqrt{\mu_\infty^2 + 6.178 \, \rho \tau_0 \, D^2(1 - \alpha)^2}}{\rho \, D(1 - \alpha)}$$

Parallel-Plate Flow, Power Law

$$\bar{v} = \frac{24.51\ Q}{D^2(1-\alpha^2)} \qquad (4-4)$$

$$\Delta P_L = \frac{L\ k}{300\ D(1-\alpha)} \left[\frac{2.4\ \bar{v}}{D(1-\alpha)} * \frac{(2n+1)}{3n} \right]^n \qquad (7-19)$$

$$Re_c = \frac{3}{2}\ (3470 - 1370\ n)$$

$$\bar{\mu} = \frac{60000\ \Delta P_L D^2 (1-\alpha)^2}{L\ \bar{v}} \qquad (7-32)$$

$$Re = \frac{15.47\ \rho\ \bar{v}\ D(1-\alpha)}{\bar{\mu}} \qquad (7-30)$$

$$\Delta P = \Delta P_L \qquad (Re \le Re_c)$$

$$y = (\log n + 3.93) / 50 \qquad (5-37)$$

$$z = (1.75 - \log n) / 7 \qquad (5-38)$$

$$f = y * Re^{-z} \qquad (Re \ge Re_c + 1200) \qquad (5-36)$$

$$f = \frac{24}{Re} + \frac{(Re - Re_c)}{1200}\ (y * Re^{-z} - \frac{24}{Re}) \qquad (Re_c < Re < [Re_c + 1200])$$

$$\Delta P = \Delta P_T = \frac{f\ L\ \rho\ \bar{v}^2}{92903\ D(1-\alpha)} \qquad (Re > Re_c) \qquad (5-32)$$

$$v_c = \left(\frac{12.88\ k * Re_c}{\rho} \right)^{\frac{1}{(2-n)}} * \left[\frac{2.4}{D(1-\alpha)} * \frac{(2n+1)}{3n} \right]^{\frac{n}{(2-n)}}$$

Annular Flow, Power Law

$$\bar{v} = \frac{24.51\ Q}{D^2(1 - \alpha^2)} \quad (4-4)$$

$$y = .37\ n^{-.14} \quad (7-26)$$

$$z = 1 - (1 - \alpha^y)^{1/y} \quad (7-27)$$

$$G = \left(1 + \frac{z}{2}\right)\left[\frac{n(3 - z) + 1}{(4 - z)n}\right] \quad (7-28)$$

$$\Delta P_L = \frac{k\ L}{300\ D(1 - \alpha)}\left[\frac{1.6\ \bar{v}\ G}{D(1 - \alpha)}\right]^n \quad (7-29)$$

$$G_N = \frac{1 + \alpha^2 + \frac{(1 - \alpha^2)}{\ln \alpha}}{(1 - \alpha)^2}$$

$$\bar{\mu} = \frac{90000\ \Delta P_L D^2 (1-\alpha)^2\ G_N}{L\ \bar{v}} \quad (7-31)$$

$$Re = \frac{15.47\ \rho\ \bar{v}\ D(1 - \alpha)}{\bar{\mu}} \quad (7-30)$$

$$Re_c = \frac{(3470 - 1370n)}{G_N}$$

$$\Delta P = \Delta P_L \qquad (Re \leqslant Re_c)$$

$$y = \frac{(\log n + 3.93)}{50} \tag{5-37}$$

$$z = \frac{(1.75 - \log n)}{7} \tag{5-38}$$

$$f = y * Re^{-z} \qquad (Re \geq Re_c + \frac{800}{G_N}) \tag{5-36}$$

$$f = \frac{16}{Re * G_N} + (Re - Re_c) \frac{G_N}{800} (y * Re^{-z} - \frac{16}{Re * G_N}) \quad (Re_c < Re < \lfloor Re_c + \frac{800}{G_N} \rfloor)$$

$$\Delta P = \Delta P_T = \frac{fL\rho \bar{v}^2}{92903 \; D(1-\alpha)} \qquad (Re > Re_c) \tag{5-32}$$

$$v_c = \left(\frac{19.33 \; k * Re_c * G_N}{\rho} \right)^{\frac{1}{(2-n)}} * \left[\frac{1.6 \; G}{D(1-\alpha)} \right]^{\frac{n}{(2-n)}}$$

Tool Joints

$$\Delta p_j = \frac{\rho(v_2^2 - v_1^2)}{4,460,000} \tag{5-46}$$

Bit Nozzles

$$v_j = \frac{418.3 \; Q}{\Sigma \; d_j^2} \tag{6-2}$$

$$\Delta p_b = \frac{\rho v_j^2}{1120} \tag{6-5}$$

Cuttings Transport

$$Re_p = \frac{15.47 \, \rho_f v_s d_p}{\bar{\mu}} \tag{9-1}$$

$$C_D = \frac{40}{Re_p} + 1 \tag{9-7}$$

$$v_s = \left[\frac{600 \, hg(\rho_p - \rho_f)}{C_D \, \rho_f}\right]^{.5} \tag{9-3}$$

$$C_a = \frac{ROP * D_b^2}{60(v_m - v_s) \, D^2(1-\alpha^2)} \tag{9-14}$$

Hydraulic Power

$$W_b = \frac{\Delta P_b * Q}{1714} \tag{10-1}$$

Impact Force

$$F_i = \frac{\rho * Q * v_j}{1930} \tag{10-5}$$

APPENDIX D
EXAMPLE HYDRAULICS CALCULATIONS

The examples provided in Appendix D are from the Exlog EAP programs "Condensed Mud Hydraulics" and "Bit Hydraulics Optimization". Calculations are provided with either equation numbers or section and page numbers. These numbers refer to the equations or discussions in the text upon which the calculations are based.

Some of the calculated values differ slightly (less than 1%) from the program printout values. The difference is due to the effect of cumulative rounding errors by the computer program.

FAST GREENBRIAR #2
TIME: 12:01   DATE: 07/19/83

MUD HYDRAULICS ANALYSIS FOR IN GAUGE HOLE

FROM ft	9505	8505	4005	5
TO ft	10005	9505	8505	4005
Length ft	500	1000	4500	4000
HOLE				
i.d. inches	10.00	10.00	10.00	15.00
Volume bbl	48.6	97.1	437.1	874.3
Volume/ft	.09714	.09714	.09714	.21857
PIPE				
i.d. inches	3.00	3.00	4.00	4.00
o.d. inches	9.00	5.00	5.00	5.00
j.d. inches	9.00	6.00	6.00	6.00
Capacity bbl	4.4	8.7	69.9	62.2
Cap/ft	.00874	.00874	.01554	.01554
Disp. bbl	35.0	16.1	41.7	37.0
Disp/ft	.06994	.01606	.00926	.00926
Cap.+Disp. bbl	39.3	24.8	111.6	99.2
Cap.+Disp./ft	.07868	.02480	.02480	.02480
ANNULUS				
Capacity bbl	9.2	72.3	325.5	775.1
Cap/ft	.01846	.07234	.07234	.19376
DOWNHOLE FLOW				
Velocity ft/min (a)	1089.4	1089.4	612.8	612.8
ANNULAR FLOW				
Velocity ft/sec (b)	8.60	2.18	2.18	.82
Velocity ft/min	516.0	130.7	130.7	49.0
Time min (c)	1.0	7.6	34.4	81.6
Slip ft/min	37.8	19.2	19.2	16.5
Transport ft/min	478.2	111.5	111.5	32.5
Transport Time min (d)	1.0	9.0	40.3	122.9
DOWNHOLE PRESSURE LOSS				
Press Drop psi B (e)	96.5	192.9	218.2	194.0
Press Drop psi P (f)	109.5	219.0	269.5	239.5
ANNULAR PRESSURE LOSS				
Flow Regime B (g)	Turbulent	Laminar	Laminar	Laminar
Flow Regime P (h)	Transition	Laminar	Laminar	Laminar
Press Drop psi B (i)	93.9	9.3	41.9	16.5
Press Drop psi P (j)	103.6	3.9	17.5	2.3
psi/100ft B	18.784	.931	.931	.412
psi/100ft P	20.711	.389	.389	.057
Critical Velocity ft/min B (k)	298.1	196.6	196.6	186.1
Critical Velocity ft/min P (l)	482.1	192.3	192.3	128.6
Reynolds Number B (m)	4066	950	950	151
Reynolds Number P (n)	2702	1514	1514	.724

B: Bingham   P: Power Law

Figure D-1: Condensed Mud Hydraulics

EAST GREENBRIAR #2
TIME: 12:05   DATE: 01/19/83

MUD HYDRAULICS ANALYSIS FOR IN GAUGE HOLE

INPUT DATA:
DEPTH	10005	ft
VERTICAL DEPTH	10005	ft
MUD FLOW RATE	400	gal/min
PUMP # 1 CAPACITY	5.400	gal/stk
DRILL RATE	200	ft/hr
CUTTINGS S.G.	2.20	g/cc
CUTTINGS SIZE	.2	inches

VOLUMES:
HOLE	1457.1	bbl
ANNULUS	1182.3	bbl
PIPE CAPACITY	145.2	bbl
PIPE DISPLACEMENT	129.7	bbl
PIPE DISP. + CAP.	275.0	bbl

LAG INFORMATION:
	MINUTES	STROKES
DOWN PIPE	15.2	1220
MUD CYCLE	139.4	11151
UP ANNULUS...		
MUD	124.1	9931
CUTTINGS	173.3	13862

MUD PROPERTIES:
DENSITY	10.0	lb/gal
PLASTIC VISCOSITY	15.00	cP
YIELD POINT	8.00	lb/cft^2
(a) MID RANGE POWER K	.268	
MID RANGE POWER N	.724	
(b) LOW RANGE POWER K	.268	
LOW RANGE POWER N	.724	

HYDRAULICS:
(c) HYDROSTATIC DEPTH	10000	ft
HYDROSTATIC PRESSURE	5190	psi
(d) JETS	10,11,12	1/32 in
JET VELOCITY	458.4	ft/sec
(e) JET PRESSURE DROP	1000	psi
(f) HYDRAULIC POWER	438.0	hp
(g) IMPACT FORCE	948.9	lbs

TOTALS:
	BINGHAM	POWER LAW
(h) SURFACE LOSS	21	21 psi
PIPE INTERNAL LOSS	702	838 psi
BIT LOSS	1800	1830 psi
ANNULAR LOSS	162	127 psi
TOTAL PRESSURE LOSS	2764	2866 psi
(i) % TOT. PRESS LOSS AT BIT	68	66
CIRCULATING PRESS AT BIT	5352	5317 psi
(j) E.C.D.	10.3	10.2 lb/gal
(k) E.C.D. W/ CUTTINGS	10.6	10.5 lb/gal
(l) FOR MUD WT	10.0	10.0 lb/gal

Figure D-2 Mud Hydraulics (cont.)

                                          EAST GREENBRIAR #2
                                          12:06   7/19/83

BIT HYDRAULICS OPTIMIZATION

 OR MUD DENSITY =   10.0 lb/gal
 REQUIRED PUMP PRESSURE =   2000 psi

 LIMITS: MAXIMUM FLOW =   374 gal/min ─────(a)
                MINIMUM FLOW =   204 gal/min ─────(b)

 FOR OPTIMUM HYDRAULIC POWER:

        OPTIMUM PRESSURE DROP AT BIT =   1247 psi ──────(c)
                   OPTIMUM PUMP FLOW =   355 gal/min ──(d)
                     RECOMMENDED FLOW =   355 gal/min ──(e)
                PRESSURE DROP AT BIT =   1247 psi ──────(f)
                  BIT HYDRAULIC POWER =   257.9 hp ──────(g)
                    JET IMPACT FORCE =   684.8 lbs ──────(h)
       RECOMMENDED TOTAL NOZZLE AREA =   .30 in^2 ─────(i)
                             NOZZLES =   11:11:12 OR EQUIVALENT ─(j)

 FOR OPTIMUM JET IMPACT FORCE:

        OPTIMUM PRESSURE DROP AT BIT =   906 psi ──────(k)
                   OPTIMUM PUMP FLOW =   444 gal/min ──(l)

        OPTIMUM FLOW ABOVE MAXIMUM ALLOWED PUMP FLOW ─────(m)
                     RECOMMENDED FLOW =   374 gal/min
                PRESSURE DROP AT BIT =   1177 psi ──────(n)
                  BIT HYDRAULIC POWER =   256.8 hp ──────(o)
                    JET IMPACT FORCE =   702.0 lbs ──────(p)
       RECOMMENDED TOTAL NOZZLE AREA =   .33 in^2 ─────(q)
                             NOZZLES =   12:12:12 OR EQUIVALENT ─(r)

Figure D-3: Bit Hydraulics Optimization

Figure D-1 - Condensed Mud Hydraulics

Line a

$$\bar{v} = \frac{24.51\ Q}{D^2} \qquad (4-4)$$

$$\frac{24.51 * 400}{3^2} = \underline{1,089.3}$$

$$\frac{24.51 * 400}{4^2} = \underline{612.8}$$

Line b

$$\bar{v} = \frac{24.51\ Q}{(D^2 - d^2)} \qquad (4-4)$$

$$\frac{24.51 * 400}{(10^2 - 9^2)} = \underline{516.0}$$

$$\frac{24.51 * 400}{(10^2 - 5^2)} = \underline{130.7}$$

$$\frac{24.51 * 400}{(15^2 - 5^2)} = \underline{49.0}$$

Line c

For turbulent flow,

$$C_D = 1.5 \tag{9-6}$$

$$v_s = 113.4 * \left[\frac{d_p (\rho_p - \rho_f)}{C_D * \rho_f}\right]^{1/2} \tag{9-3}$$

$$113.4 \left[\frac{.2(8.33 * 2.2 - 10)}{1.5 * 10}\right]^{1/2} = \underline{37.8}$$

For laminar flow,

$$\bar{\mu} = \frac{\Delta P * 60,000 (D - d)^2}{L * \bar{v}} \tag{7-32}$$

$$v_s = 175 * d_c \left[\frac{(\rho_p - \rho_f)^2}{\bar{\mu} * \rho_f}\right]^{1/3} \qquad \begin{matrix}(9-1)\\(9-2)\\(9-5)\end{matrix}$$

$$\frac{.389}{100} * \frac{60,000}{130.7} * (10 - 5)^2 = 44.64$$

$$175.2 * 2 \left[\frac{(833 * 2.2 - 10)^2}{44.64 * 10}\right]^{1/3} = \underline{18.8}$$

$$\frac{.057}{100} * \frac{60{,}000}{49.0} * (15 - 5)^2 = 69.80$$

$$175.2 * 2 \left[\frac{(8.33 * 2.2 - 10)^2}{69.80 * 10}\right]^{1/3} = \underline{16.2}$$

Line d

$$v_L = v_m - v_s \qquad (9\text{-}11)$$

$$516.0 - 37.8 \qquad = \underline{478.2}$$

$$130.7 - 19.2 \qquad = \underline{111.5}$$

$$49.0 - 16.5 \qquad = \underline{32.5}$$

Line e

In calculating Reynolds number for Bingham fluids, the program uses an average viscosity of PV/3.2. This formula is cited in Moore.

$$Re = \frac{15.47 \, \rho \, \bar{v} \, D}{(PV/3.2)} \qquad (5\text{-}27)$$

$$f = \frac{.046}{Re^{.20}} \qquad (5\text{-}36)$$

$$\Delta P = \frac{f \, L \, \rho \, \bar{v}^2}{92{,}915 * D} \qquad (5\text{-}32)$$

$$Re = \frac{15.47 * 10 * 1{,}089.4 * 3}{(15/3.2)} = 107{,}859.0$$

$$f = \frac{.046}{107,859^{.2}} = .0045$$

$$\Delta P = \frac{.0045 * 500 * 10 * 1,089.4^2}{92,915 * 3} = \underline{96.5}$$

$$\Delta P = \frac{.0045 * 1,000 * 10 * 1,089.4^2}{92,915 * 3} = \underline{192.9}$$

$$Re = \frac{15.47 * 10 * 612.8 * 4}{(15/3.2)} = 80,896$$

$$f = \frac{.046}{80,896^{.2}} = .0048$$

$$\Delta P = \frac{.0048 * 4,500 * 10 * 612.8^2}{92,915 * 4} = \underline{218.2}$$

$$\Delta P = \frac{.0048 * 4,000 * 10 * 612.8^2}{92,915 * 4} = \underline{194.0}$$

Line f

$$\Delta P_L = \frac{L * k}{300 \, D} * \left[\frac{1.6 * \overline{v}}{D} * \frac{(3n + 1)}{4n}\right]^n \tag{5-18}$$

$$\overline{\mu} = \frac{90,000 \, \Delta P_L * D^2}{L * \overline{v}} \tag{5-28}$$

$$Re = \frac{15.47 * \rho * \bar{v} * D}{\bar{\mu}} \qquad (5\text{-}27)$$

$$f = \frac{(\log n + 3.93)}{50} * Re \uparrow \left(\frac{\log n - 1.75}{7}\right) \qquad \begin{matrix}(5\text{-}36)\\(5\text{-}37)\\(5\text{-}38)\end{matrix}$$

$$\Delta P_T = \frac{f\,L\,\rho\,\bar{v}^2}{92{,}894 * D} \qquad (5\text{-}32)$$

$$\Delta P_L = \frac{500 * .268}{300 * 3} * \left[\frac{1.6 * 1{,}089.4}{3} * \frac{(3 * .724 + 1)}{4 * .724}\right]^{.724} = 15.95$$

$$\bar{\mu} = \frac{90{,}000 * 15.95 * 3^2}{500 * 1{,}089.4} = 23.72$$

$$Re = \frac{15.47 * 10 * 1{,}089.4 * 3}{23.72} = 21{,}316$$

$$f = \frac{(\log .724 + 3.93)}{50} * 21{,}316 \uparrow \left(\frac{\log .724 - 1.75}{7}\right) = .0051$$

$$\Delta P_T = \frac{.0051 * 500 * 10 * 1{,}089.4^2}{92{,}894 * 3} = \underline{109.4}$$

$$\Delta P_T = \frac{.0051 * 1000 * 10 * 1{,}089.4^2}{92{,}894 * 3} = \underline{218.8}$$

$$\Delta P_L = \frac{4{,}500 * .268}{300 * 4} * \left[\frac{1.6 * 612.8}{4} * \frac{(3 * .724 + 1)}{4 * .724}\right]^{.724} = 57.64$$

$$\bar{\mu} = \frac{90{,}000 * 57.64 * 4^2}{4{,}500 * 612.8} = 30.10$$

$$Re = \frac{15.47 * 10 * 612.8 * 4}{30.10} = 12{,}599$$

$$f = \frac{(\log .724 + 3.93)}{50} * 12{,}599 + \left(\frac{\log .724 - 1.75}{7}\right) = .0059$$

$$\Delta P_T = \frac{.0059 * 4{,}500 * 10 * 612.8^2}{92{,}894 * 4} = \underline{269.3}$$

$$\Delta P_T = \frac{.0059 * 4{,}000 * 10 * 612.8^2}{92{,}894 * 4} = \underline{239.4}$$

Line g

The flow regime, laminar or turbulent, is determined by comparing the value in line **b** with that in line **k**.

line h

$$Re_L = 3470 - 1370n \qquad \text{Paragraph heading 5.9, Paragraph 3}$$

$$Re = \frac{15.47 * \rho * \bar{v} * D}{\bar{\mu}} \qquad (5\text{-}27)$$

$$\qquad (5\text{-}36)$$
$$f = \frac{(\log n + 3.93)}{50} * Re \uparrow \left(\frac{\log n - 1.75}{7}\right) \qquad (5\text{-}37)$$
$$\qquad (5\text{-}38)$$

$$\Delta P_T = \frac{f \, L \, \rho \, \bar{v}^2}{92{,}894 * D} \qquad (5\text{-}32)$$

$$\Delta P_L = \frac{500 * .268}{300 * 3} * \left[\frac{1.6 * 1{,}089.4}{3} * \frac{(3 * .724 + 1)}{4 * .724}\right]^{.724} = 15.95$$

$$\bar{\mu} = \frac{90{,}000 * 15.95 * 3^2}{500 * 1{,}089.4} = 23.72$$

$$Re = \frac{15.47 * 10 * 1{,}089.4 * 3}{23.72} = 21{,}316$$

$$f = \frac{(\log .724 + 3.93)}{50} * 21{,}316 \uparrow \left(\frac{\log .724 - 1.75}{7}\right) = .0051$$

$$\Delta P_T = \frac{.0051 * 500 * 10 * 1{,}089.4^2}{92{,}894 * 3} = \underline{109.4}$$

$$\Delta P_T = \frac{.0051 * 1000 * 10 * 1,089.4^2}{92,894 * 3} = \underline{218.8}$$

$$\Delta P_L = \frac{4,500 * .268}{300 * 4} * \left[\frac{1.6 * 612.8}{4} * \frac{(3 * .724 + 1)}{4 * .724}\right]^{.724} = 57.64$$

$$\bar{\mu} = \frac{90,000 * 57.64 * 4^2}{4,500 * 612.8} = 30.10$$

$$Re = \frac{15.47 * 10 * 612.8 * 4}{30.10} = 12,599$$

$$f = \frac{(\log .724 + 3.93)}{50} * 12,599 + \left(\frac{\log .724 - 1.75}{7}\right) = .0059$$

$$\Delta P_T = \frac{.0059 * 4,500 * 10 * 612.8^2}{92,894 * 4} = \underline{269.3}$$

$$\Delta P_T = \frac{.0059 * 4,000 * 10 * 612.8^2}{92,894 * 4} = \underline{239.4}$$

Line g

The flow regime, laminar or turbulent, is determined by comparing the value in line **b** with that in line **k**.

line h

$$Re_L = 3470 - 1370n \qquad \text{Paragraph heading 5.9, Paragraph 3}$$

$$Re_T = 4270 - 1370n \qquad \text{Paragraph heading 5.9}$$
Paragraph 3

$$3470 - 1370 * .724 = 2478 \text{ (laminar/transitional)}$$

$$4270 - 1370 * .724 = 3278 \text{ (transitional/turbulent)}$$

The flow regime is determined by comparing the value in line n with the above critical values.

Line i

For turbulent flow,

$$Re = \frac{15.47 \, \rho \, \bar{v} \, (D - d)}{(PV / 3.2)} \qquad (7-30)$$

$$f = \frac{.046}{Re^{.2}} \qquad (5-36)$$

$$\Delta P = \frac{f L \rho \bar{v}^2}{92,915 \, (D - d)} \qquad (5-32)$$

$$Re = \frac{15.47 * 10 * 516.0 \, (10 - 9)}{(15 / 3.2)} = 17,029$$

$$f = \frac{.046}{17,029^{.2}} = .0066$$

$$\Delta P = \frac{.0066 * 500 * 10 * 516.0^2}{92,915 \; (10 - 9)} \qquad = \underline{93.9}$$

For laminar flow,

$$\Delta P = \frac{L * YP}{200 \; (D - d)} + \frac{L * PV * \bar{v}}{60,000 \; (D - d)^2} \qquad (7\text{-}25)$$

$$\frac{1000 * 8}{200 \; (10 - 5)} + \frac{1000 * 15 * 130.7}{60,000 \; (10 - 5)^2} \qquad = \underline{9.3}$$

$$\frac{4500 * 8}{200 \; (10 - 5)} + \frac{4500 * 15 * 130.7}{60,000 \; (10 - 5)^2} \qquad = \underline{41.9}$$

$$\frac{4000 * 8}{200 \; (15 - 5)} + \frac{4000 * 15 * 49.0}{60,000 \; (15 - 5)^2} \qquad = \underline{16.5}$$

line j

For transitional flow,

$$f = \frac{16}{Re_c} + \frac{(Re - Re_c)}{800} * \left[\frac{(3.93 + \log n)}{50} * Re_c \uparrow \left(\frac{\log n - 1.75}{7} \right) - \frac{16}{Re_c} \right]$$

This expression differs slightly from the corresponding expression for friction factor given on page C-21 of Appendix C.

$$\Delta P_T = \frac{f \, L \, \rho \, \bar{v}^2}{92,894 \, (D - d)} \tag{5-32}$$

$$f = \frac{16}{2478} + \frac{(2702 - 2478)}{800} \left[\frac{(3.93 + \log .724)}{50} * 2478 \uparrow \left(\frac{\log .724 - 1.75}{7} \right) - \frac{16}{2478} \right] = .$$

$$\Delta P_T = \frac{.0072 * 500 * 10 * 516.0^2}{92,894 \, (10 - 9)} \qquad = \underline{103.5}$$

For laminar flow,

$$x = .37 \, n^{-.14} \tag{7-26}$$

$$z = 1 - \left[1 - \left(\frac{d}{D} \right)^x \right]^{1/x} \tag{7-27}$$

$$G = \left[\frac{(3 - z) \, n + 1}{(4 - z) \, n} \right] * \left(1 + \frac{z}{2} \right) \tag{7-28}$$

$$\Delta P = \frac{k \, L}{300 \, (D - d)} \left[\frac{1.6 \, \bar{v} \, G}{(D - d)} \right]^n \tag{7-29}$$

$$x = .37 * .724^{-.14} = .3871$$

$$z = 1 - \left[1 - \left(\frac{5}{10}\right)^{.3871}\right]^{1/.3871} = .9762$$

$$G = \left[\frac{(3 - .9762) * .724 + 1}{(4 - .9762) * .724}\right] * \left(1 + \frac{.9762}{2}\right) = 1.6757$$

$$\Delta P = \frac{.268 * 1000}{(10 - 5)} \left[\frac{1.6 * 130.7 * 1.6757}{(10 - 5)}\right]^{.724} = \underline{3.9}$$

$$\Delta P = \frac{.268 * 4500}{300 (10 - 5)} \left[\frac{1.6 * 130.7 * 1.6757}{(10 - 5)}\right]^{.724} = \underline{17.4}$$

$$z = 1 - \left[1 - \left(\frac{5}{15}\right)^{.3871}\right]^{1/.3871} = .9353$$

$$G = \left[\frac{(3 - .9353) * .724 + 1}{(4 - .9353) * .724}\right] * \left(1 + \frac{.9353}{2}\right) = 1.6502$$

$$\Delta P = \frac{.268 * 4000}{300 (15 - 5)} \left[\frac{1.6 * 49.0 * 1.6502}{(15 - 5)}\right]^{.724} = \underline{2.3}$$

Line k

$$v_c = \frac{64.68 \, PV + 64.68 \left[PV^2 + 9.271 \, YP * \rho \, (D - d)^2\right]^{1/2}}{\rho * (D - d)}$$

This equation is analogous to the equation discussed under paragraph heading 7.12, paragraph 3, but uses a critical Reynolds number of 2000.

$$v_c = \frac{64.68 * 15 + 64.68 \left[15^2 + 9.271 * 8 * 10 (10 - 9)^2\right]^{1/2}}{10 (10 - 9)}$$

$$= \underline{298.1}$$

$$v_c = \frac{64.68 * 15 + 64.68 \left[15^2 + 9.271 * 8 * 10 (10 - 5)^2\right]^{1/2}}{10 (10 - 5)}$$

$$= \underline{196.6}$$

$$v_c = \frac{64.68 * 15 + 64.68 \left[15^2 + 9.271 * 8 * 10 (15 - 5)^2\right]^{1/2}}{10 (15 - 5)}$$

$$= \underline{186.1}$$

Line 1

$$v_c = 60 \left[\frac{Re_L * k}{185.6 * \rho} * \left(\frac{96 * G}{(D - d)}\right)^n \right]^{\frac{1}{(2 - n)}}$$

This equation is analogous to the equation discussed under paragraph heading 7.12, paragraph 3, but with constant terms regrouped.

$$z = 1 - \left[1 - \left(\frac{9}{10}\right)^{.3871}\right]^{1/.3871} = .9998$$

$$G = \left[\frac{(3 - .9998) * .724 + 1}{(4 - .9998) * .724}\right] * \left(1 + \frac{.9998}{2}\right) = 1.6905$$

$$v_c = 60\left[\frac{2478 * .268}{185.6 * 10} * \left(\frac{96 * 1.6905}{(10 - 9)}\right)^{.724}\right]^{\frac{1}{(2 - .724)}} = \underline{481.4}$$

$$v_c = 60\left[\frac{2478 * .268}{185.6 * 10} * \left(\frac{96 * 1.6757}{(10 - 5)}\right)^{.724}\right]^{\frac{1}{(2 - .724)}} = \underline{192.2}$$

$$v_c = 60\left[\frac{2478 * .268}{185.6 * 10} * \left(\frac{96 * 1.6502}{(10 - 5)}\right)^{.724}\right]^{\frac{1}{(2 - .724)}} = \underline{128.6}$$

Line m

$$\Delta P_L = \frac{L * YP}{200 (D - d)} + \frac{L * PV * \overline{v}}{60,000 (D - d)^2} \qquad (7\text{-}25)$$

$$\overline{\mu} = \frac{60,000 \, \Delta P_L (D - d)^2}{L * \overline{v}} \qquad (7\text{-}32)$$

$$Re = \frac{15.47 \, \rho \, \overline{v} \, (D - d)}{\overline{\mu}} \qquad (7\text{-}30)$$

$$\Delta P_L = \frac{500 * 8}{200 (10 - 9)} + \frac{500 * 15 * 516.0}{60,000 (10 - 9)^2} = 84.50$$

$$\overline{\mu} = \frac{60000 * 84.50 * (10 - 9)^2}{500 * 516.0} = 19.65$$

$$Re = \frac{15.47 * 10 * 516.0 \ (10-9)}{19.65} = \underline{4062}$$

$$\Delta P_L = \frac{4500 * 8}{200 \ (10-5)} + \frac{4500 * 15 * 130.7}{60,000 \ (10-5)^2} = 41.88$$

$$\bar{\mu} = \frac{60,000 * 41.88 \ (10-5)^2}{4500 * 130.7} = 106.8$$

$$Re = \frac{15.47 * 10 * 130.7 \ (10-5)}{106.8} = \underline{946}$$

$$\Delta P_L = \frac{4000 * 8}{200 \ (15-5)} + \frac{4000 * 15 * 49.0}{60,000 \ (15-5)^2} = 16.49$$

$$\bar{\mu} = \frac{60,000 * 16.49 * (15-5)^2}{4000 * 49.0} = 504.8$$

$$Re = \frac{15.47 * 10 * 49.0 * (15-5)}{504.8} = \underline{150}$$

Line n

$$\Delta P_L = \frac{L * k}{300 \ (D-d)} * \left[\frac{1.6 \ \bar{v} \ G}{(D-d)}\right]^n \tag{7-29}$$

$$\bar{\mu} = \frac{90,00 \ \Delta P_L \ D^2 (1-\alpha)^2}{L * \bar{v}} \tag{7-31}$$

$$Re = \frac{15.47 * \rho * \overline{v} (D - d)}{\overline{\mu}} \qquad (7\text{-}30)$$

$$\Delta P_L = \frac{500 * .268}{300 (10 - 9)} * \left[\frac{1.6 * 516.0 * 1.6905}{(10 - 9)}\right]^{.724} = 84.49$$

$$\overline{\mu} = \frac{90,000 * 84.49 * (10 - 9)^2}{500 * 516.0} = 29.47$$

$$Re = \frac{15.47 * 10 * 516.0 * (10 - 9)}{29.47} = \underline{2,708}$$

$$\overline{\mu} = \frac{90,000 * 17.4 * (10 - 5)^2}{4500 * 130.7} = 66.56$$

$$Re = \frac{15.47 * 10 * 130.7 (10 - 5)}{66.56} = \underline{1519}$$

$$\overline{\mu} = \frac{90,000 * 2.3 * (15 - 5)^2}{4000 * 49.0} = 105.6$$

$$Re = \frac{15.47 * 10 * 49.0 (15 - 5)}{105.6} = \underline{718}$$

Figure D-2 - Condensed Mud Hydraulics (cont.)

Line a

$$k = 1.067 * \frac{(PV + YP)}{511^n} \qquad (2-41)$$

$$1.067 * \frac{(15 + 8)}{511^{.714}} = \underline{0.268}$$

Line b

$$n = \log \left[\frac{(2PV + YP)}{(PV + YP)}\right] / \log (2) \qquad (2-29)$$

$$\log \left[\frac{(2 * 15 + 8)}{(15 + 8)}\right] / \log (2) = \underline{0.724}$$

Line c

$$P_h = .0519 * \rho * D_v \qquad (1-1)$$

$$.0519 * 10 * 10,000 = \underline{5190.0}$$

Line d

$$v_j = \frac{418.3 \; Q}{\Sigma dj^2} \qquad (6-2)$$

$$\frac{418.3 * 400}{(10^2 + 11^2 + 12^2)} = \underline{458.4}$$

Line e

$$\Delta P_b = \frac{\rho * v_j^2}{1120} \tag{6-5}$$

$$\frac{10 * 458.4^2}{1120} = \underline{1876.0}$$

Line f

$$W_b = \frac{\Delta P_b * Q}{1714} \tag{10-1}$$

$$\frac{1876 * 400}{1714} = \underline{438.7}$$

Line g

$$F_i = \frac{\rho * Q * v_j}{1932} \tag{10-5}$$

$$\frac{10 * 400 * 458.4}{1932} = \underline{949.1}$$

Line h

$$P_s = 10^{-5} * k_s * \rho * Q^{1.86}$$

$$10^{-5} * 3 * 10 * 400^{1.86} = \underline{21}$$

Line i

% Total Pressure Loss at Bit = $100 * \dfrac{\text{Bit Loss}}{\text{Total Pressure Loss}}$

$$100 * \dfrac{1880}{2764} = \underline{68}$$

$$100 * \dfrac{1880}{2866} = \underline{66}$$

Line j

Circulating Pressure at Bit = Hydrostatic Pressure + Annular Loss

$$5190 + 162 = \underline{5352}$$

$$5190 + 127 = \underline{5317}$$

Line k

$$ECD = \rho_f + \dfrac{\Sigma P_a}{.0519 * D_v} \qquad (4\text{-}6)$$

$$10 + \dfrac{162}{.0519 + 10{,}000} = \underline{10.3}$$

$$10 + \dfrac{127}{.0519 * 10{,}000} = \underline{10.2}$$

Line 1

At * v_s = 0

$$C_a = \frac{ROP * D_b^2}{60 \bar{v} (D^2 - d^2)} \qquad (9-14)$$

$$\therefore C_a = \frac{ROP \, D_b^2}{60 * Q * 24.51} \qquad (9-17)$$

$$\frac{200 * 10^2}{60 * 400 * 24.51} = 0.0340$$

$$\Delta P_c = .0519 * C_a (\rho_p - \rho_f) * D_v \qquad (9-18)$$

$$.0519 * .0340 (2.2 * 8.33 - 10) * 16,000 = 147$$

$$ECD_c = \rho + \frac{(\Sigma P + \Delta P_c)}{.0519 \, D_v} \qquad (9-20)$$

$$10 + \frac{(162 + 147)}{.0519 * 10,000} = \underline{10.6}$$

$$10 + \frac{(127 + 147)}{.0519 * 10,000} = \underline{10.5}$$

Figure D-3 - Bit Hydraulics Optimization

Line a

The criterion for maximum flowrate is critical velocity opposite collars.

$$Q(max) = Q * \frac{v_c}{v}$$

$$400 * \frac{482.1}{516.0} = \underline{374.0}$$

Line b

Minimum flowrate is chosen to give 24 ft/min in the top annular section.

$$Q(min) = Q * \frac{v(min)}{v}$$

$$400 * \frac{25}{49.0} = \underline{204.0}$$

Line c

As slow circulating data is not provided, a default is used for the parasitic pressure exponent. The following default value is consistent with Power Law turbulent flow calculations:

$$m = 2 + (2 - n) * \frac{(\log n - 1.75)}{7}$$

$$2 + (2 - .724) * \frac{(\log .724 - 1.75)}{7} = 1.6554$$

$$P_p(opt) = P_s(max)/(m + 1) \qquad (10\text{-}3)$$

$$P_b(opt) = P_s(max) - P_p(opt)$$ Section 10, Paragraph 6

$$P_p(opt) = \frac{2000}{(1.6554 + 1)} = 753$$

$$P_b(opt) = 2000 - 753 \qquad = \underline{1247}$$

Line d

For the previous bit run,

$$P_p = P_s - P_b$$ Section 10, Paragraph 6

$$2800 - 1880 = 920$$

Noting that for any drilling assembly, the parasitic pressure is proportional to the m power of flowrate,

$$\frac{P_{p1}}{P_{p2}} = \left(\frac{Q_1}{Q_2}\right)^m$$

Therefore,

$$Q_2 = Q_1 \left(\frac{P_{p2}}{P_{p1}}\right)^{1/m}$$

The parasitic pressure P_{p2} is the difference between the desired maximum pressure, which is specified, and the optimum bit pressure, which was determined in the previous calculation. Thus,

$$Q(opt) = Q * \left[\frac{\{P_s(max) - P_b(opt)\}}{P_p}\right]^{1/m}$$

$$400 * \left[\frac{(2000 - 1247)}{920}\right]^{1/1.6554} \qquad = \underline{354}$$

Line e

Optimum flow is between minimum and maximum limits.

Line f

The optimum pressure from line c is retained.

Line g

$$W_b = \frac{P_b * Q(opt)}{1714} \tag{10-1}$$

$$\frac{1247 * 355}{1714} = \underline{258}$$

Line h

$$v_j = \left[\frac{1120 * P_b}{\rho}\right]^{1/2} \tag{6-5}$$

$$F_i = \frac{\rho * Q(opt) * v_j}{1932} \tag{10-5}$$

$$v_j = \left[\frac{1120 * 1247}{10}\right]^{1/2} = 374$$

$$F_i = \frac{10 * 355 * 374}{1932} = \underline{687}$$

Line i

$$\Sigma d_j^2 = \frac{418.3 * Q(opt)}{v_j} \tag{6-2}$$

$$A_n = \frac{\pi * \Sigma d_j^2}{4096} \tag{6-2}$$

$$\Sigma d_j^2 = \frac{418.3 * 355}{374} = 397$$

$$A_n = \frac{\pi * 397}{4096} \qquad = \underline{0.30}$$

Line j

$$A_n = \frac{\pi}{4096} (\underline{12^2} + \underline{11^2} + \underline{11^2}) = 0.30$$

Any combination of nozzle sizes may be used as long as total nozzle area equals 0.30. The program distributes the recommended nozzle area as equally as possible among the specified number of nozzles.

Line k

$$P_p(opt) = 2P_s(max)/(m+2) \tag{10-7}$$

$$P_b(opt) = P_s(max) - P_p(opt) \qquad \text{Section 10, Paragraph 6}$$

$$P_p(opt) = \frac{2 * 2000}{(1.6554 + 2)} = 1094$$

$$P_b(opt) = 2000 - 1094 \qquad = \underline{906}$$

Line 1

$$Q(opt) = Q * \left[\frac{\{P_s(max) - P_b(opt)\}}{P_p}\right]^{1/m}$$

$$400 * \left[\frac{(2000 - 906)}{920}\right]^{1/1.6554} = \underline{444}$$

Line m

Since optimum flowrate exceeds maximum pump flow, the maximum permissible flow is used.

Line n

$$P_b = P_s(max) - P_p * \left(\frac{Q(max)}{Q}\right)^m$$

$$2000 - 920 * \left(\frac{374}{400}\right)^{1.6554} = \underline{1177}$$

Line o

$$W_b = \frac{P_b * Q(max)}{1714} \tag{10-1}$$

$$\frac{1177 * 374}{1714} = \underline{257}$$

Line p

$$v_j = \left[\frac{1120 \, P_b}{\rho}\right]^{1/2} \tag{6-5}$$

$$F_j = \frac{\rho * Q(max) \, v_j}{1932} \tag{10-5}$$

$$\left[\frac{1120 * 1177}{10}\right]^{1/2} = 363$$

$$\frac{10 * 374 * 363}{1932} = \underline{703}$$

Line q

$$\Sigma d_j^2 = \frac{418.3 \, Q(max)}{v_j} \tag{6-2}$$

$$A_n = \frac{\pi * \Sigma d_j^2}{4096}$$

$$\Sigma d_j^2 = \frac{418.3 * 374}{363} = 431$$

$$A_n = \frac{\pi * 431}{4096} = \underline{0.33}$$

Line r

$$A_n = \frac{\pi}{4096} * \left(\underline{12^2} + \underline{12^2} + \underline{12^2}\right) = 0.33$$

Any combination of nozzle sizes may be used as long as total nozzle area equals 0.33. The program distributes the recommended nozzle area as equally as possible among the specified number of nozzles.

GLOSSARY

annulus: The space between two cylinders, one of which is contained within the other. Usually refers to the space between the drillstring and the wellbore.

apparent viscosity: Shear stress divided by shear rate.

attapulgite: Clay mineral commonly used to provide viscosity in salty muds. Attapulgite particles are acicular, while those of most other clays used in drilling fluids are platy.

bentonite: A naturally occurring montmorillonite clay that is commonly used as a viscosifier in fresh-water muds.

Bernoulli's principle: A formula relating fluid pressure to velocity, which can be expressed in the form:

hydrostatic pressure = flowing pressure + dynamic pressure. This formula neglects the effects of fluid friction.

Bingham number: A dimensionless quantity used to indicate the degree of departure from Newtonian behavior of a Bingham fluid. It is calculated from yield point * diameter/ plastic viscosity / bulk velocity.

Bingham plastic: A fluid model represented by the relationship: shear stress = yield point + (plastic viscosity * shear rate).

Blasius formula: A formula relating Reynolds number and turbulent friction factor as a power function, or a straight line on a log-log plot.

booster pump: A pump used to increase fluid velocity from the BOPs to the flowline in long risers.

bulk velocity: Flowrate divided by a cross-sectional area of flowpath.

Casson fluid: A two-parameter fluid model representing behavior intermediate between Bingham and Power Law.

clinging constant: A parameter used in calculating swab and surge pressures. It gives a convenient way to find the effective mud velocity that would develop the required pressure drop in a stationary annulus.

consistency factor: One of the two parameters characterizing a Power Law fluid.

Couette flow: Flow between two concentric cylinders which rotate with respect to each other, as in a rotary viscometer.

critical velocity: The bulk velocity above which laminar flow ceases.

dilatant fluid: A fluid whose apparent viscosity increases as shear rate is increased.

double-acting pump: A pump which displaces fluid when the piston travels in either direction.

drag coefficient: A dimensionless number representing the ratio of the stress tending to oppose motion to the dynamic fluid pressure.

duplex pump: A reciprocating pump with the cylinders.

dynamic pressure: A property of fluid in motion equal to the kinetic energy of unit volume of fluid.

Fanning friction factor: In turbulent flow, the ratio of shear stress opposing motion at the wall of the flow channel to dynamic pressure.

flash gels: High gel strength immediately after shearing the fluid, usually indicating flocculation.

flocculation: Aggregation of clay particles suspended in a fluid.

flow behavior index: A parameter describing Power Law fluids, usually used to indicate the degree of departure from Newtonian behavior.

flow regime: A term describing general characteristics of fluid flow. May be plug flow, or laminar, transitional, or turbulent flow.

friction factor: See "Fanning friction factor."

gel strength: Resistance to shearing of stationary fluid.

Hedstrom number: A dimensionless number characterizing a Bingham fluid.

helical flow: A combination of annular flow and Couette flow, such as occurs in the annulus outside a rotating drillstring.

Herschel-Bulkley fluid: three-parameter non-Newtonian fluid model.

hydraulic power: Power required to cause a fluid to flow: the product of flowrate and pressure drop.

hydrostatic pressure: The pressure exerted by gravity acting on a static column of fluid; the product of the vertical height of the column, the fluid's density, and the acceleration due to gravity.

impact force: The thrust developed by a moving jet of fluid: the product of dynamic pressure and the cross-sectional area of the jet.

impact pressure: Impact force where the jet of fluid from bit nozzles strikes the bottom of the hole, divided by area of impact.

initial gel: Gel strength measured 10 seconds after shearing the fluid.

Von Karman formula: A mathematical relationship between Reynolds number and Fanning friction factor.

laminar flow: Flow in which all fluid particles move along streamlines.

Marsh funnel: Instrument for testing viscosity in the field.

mean viscosity: The viscosity of the Newtonian fluid which, in laminar flow, would develop the same pressure drop as the fluid being circulated or tested.

Moineau motor: Downhole motor working on the positive-displacement principle.

montmorillonite: Clay mineral widely used to provide viscosity in water-based drilling fluids.

Moody diagram: Log-log plot of Fanning friction factor against Reynolds number.

Newtonian fluid: A fluid having viscosity independent of shear rate.

Ostwald-de Walde fluid: Power Law fluid.

parasitic pressure loss: The contribution to total pressure drop made by all parts of the circulating system excluding the bit nozzles.

plastic viscosity: In a Bingham fluid, the slope of the graph of shear stress versus shear rate. An apparent plastic viscosity is reported for drilling fluids, based on two viscometer measurements.

plasticity: See "Bingham number".

plug flow: Flow with no deformation; shear rate is zero in a plug flow region.

poiseville flow: Flow inside a circular tube.

Power Law fluid: A fluid in which shear stress is a power function of shear rate.

progressive gels: A 10-minute gel strength very much higher than the 10-second gel reading; also called strong gels.

pseudoplastic: A fluid in which apparent viscosity decreases as shear rate is increased.

Reynolds number: A dimensionless correlation parameter. For Newtonian fluids flowing in circular tubes, the Reynolds number is

Density * bulk velocity * diameter/viscosity. For non-Newtonian fluids and other geometries, there is no universally accepted definition.

rheology: The study of the flow of matter.

Robertson-Stiff fluid: A three-parameter non-Newtonian fluid model.

shear rate: The rate of deformation of an element of fluid.

shear stress: The stress developed by a fluid, tending to resist deformation.

shear thinning: A reduction in apparent viscosity as shear rate is increased.

single-acting: A pump in which fluid is discharged only on the forward stroke of the piston.

slip velocity: The velocity with which a particle sinks relative to the fluid surrounding it.

stability parameter: A dimensionless quantity proposed as a means of calculating the onset of turbulence.

Stanton diagram: A plot of friction factor against Reynolds number.

supercharging pump: A centrifugal pump feeding fluid under positive pressure to a reciprocating pump.

thixotropy: A reduction in viscosity as fluid is sheared for a period of time.

transitional flow: A flow regime in which characteristics are intermediate between laminar and turbulent flow.

transport ratio: Net upward velocity of suspended particles, divided by annular velocity.

transport velocity: Net upward velocity of particles, obtained by subtracting slip velocity from annular velocity.

triplex pump: A reciprocating pump containing three cylinders.

turbodrill: A downhole motor operating on the turbine principle.

turbulent flow: Flow with numerous eddies, in which the fluid becomes completely mixed.

upset: A portion of a pipe with increased wall thickness.

viscosity: A measure of a fluid's resistance to flow. In a Newtonian fluid, it can be defined as the ratio of shear stress to shear rate.

volumetric efficiency: The volume of fluid output by a pump during one stroke, divided by the volume swept by the pistons.

yield point: In a Bingham fluid, the stress required to initiate deformation. An apparent yield point is usually reported, based on viscometer readings at two shear rates.

yield-pseudoplastic: See "Robertson-Stiff fluid".

yield stress: The stress required to initiate deformation in a non-Newtonian fluid.

z-factor: See "stability parameter".

INDEX

Absolute roughness, 71
Absolute viscosity, 16–17
Angular velocity, 30, 32
Annular flow
   analysis of, 91–106
   Power Law equation for, 154–155, 164–165
Annular pressure loss
   approximations of, 97–100
   calculation of, 91–96
   practical methods for solving, 100–103
Annular velocity, 56
Annulus, 56, 91–106
   average fluid velocity of, 15
   fluid flow up, 7
   laminar flow in, 91–103
   transitional flow in, 105
   turbulent flow in, 103–104
Apparent viscosity, 18, 19, 22
Attapulgite, 48
Average fluid velocity, 15

Bentonite, 47–48
Bingham fluid
   in annulus, 96, 97, 100–102, 105–106
   field procedures for, 39, 66–67
   gel strength and, 42–44
   laminar flow in, 65, 66, 96, 97, 100–102
   model of, 17–18
   plasticity in, 67
   Power Law model and, 19–21
   properties of, 36–37
   transitional flow, 105–106
   shear rate in, 35
   shear stress in, 34
Bingham fluid model flow curve
   drilling fluid curve and, 20 (fig. 2-10), 21
Bingham number, 67, 100
Bingham Plastic fluid model. *See* Bingham fluid
Bit
   circulating pressure at, 187
   pressure loss at, 55–56, 86–88, 187
Bit/drive combinations, calculations for, 88
Bit hydraulics optimization, 24, 170 (fig. D-3)
Bit nozzles, 86–88
   equation for, 156, 165–166
Blasius relationship, 73, 75 (fig. 5-5)
Booster pumps, 129
Bourgoyne, A. T., 122

Calcium montmorillonite, 48
Casson fluid
   in annulus, 98, 103
   gel strength and, 42–44
   laminar flow of, 65, 98, 103
   model of, 21–22
   plasticity in, 67
   properties of, 37
   shear rate in, 35
   shear stress in, 34
Centipoise, 39

Circulating pressure, at bit, 187
Clinging constant, 114–117
Colebrook equation, 73
Consistency factor, 19, 21
   calculation of, 39–41
Continuity of flow, principle of, 14–15
Core bit, 88
Correction factor, for Power Law
   consistency factor, 40–41
Critical Reynolds number, 70, 180
Critical velocity
   critical Reynolds number and, 70
   flow regime determination and, 13–14
   Reynolds number and, 13–14
Cuttings, pressure drop due to, 126
Cuttings transport, 119–127, 156, 166
Cuttings velocity, 124–126

Deformation, fluid, 8–10
Density. *See* Fluid density
Diamond bit, 88
Dilatant fluid, 19
Dodge, D. W., 73–74, 76
Drag coefficient, 121–122
Drilling fluid, 2, 47–50
   components of, 47
   cuttings transport by, 119–127
   flow curve of, 20 (fig. 2-10), 21
   flow of, in drillstring, 59–79
   functions of, 47
   gel strength of, 23
   hydrostatic pressure of, 3–4
   non-Newtonian fluids and, 16
   Power Law fluid model and, 19–21
   shear thinning in, 19–21
   time-dependent flow behavior of, 22–23
   yield stress in, 21
Drilling hydraulics, principles of, 1–5
Drilling muds. *See* Drilling fluids
Drillstring, 55, 59–79
   average fluid velocity of, 15
   fluid flow down, 7
   swab and surge pressures and, 111–113

ECD. *See* Effective circulating density
Effective circulating density, 57, 126–127
Effective viscosity, 18
Emulsions, 47, 49
Equations
   key, 145–166
   in oilfield units, 145, 146, 157–166
   in S. I. units, 145–156
Equivalent Reynolds number, 69

Fann viscometer, 38–39
Fanning equation, 70–71
Fanning friction factor, 71–72
Field procedures
   for Bingham fluid, 39, 66–67
   for fluid flow measurements, 38–44

for laminar flow, 66–67
for Newtonian fluid, 39
for Power Law fluid, 39–44
Flocculation, 48
Flow. *See also* Fluid flow; Laminar flow; Transitional flow; Turbulent flow
  continuity of, 14–15
  defined, 7
  plug, 10
Flow behavior index, 19, 98
Flow-pressure loss, flowrate and, 19
Flow regime, 10–14
  determination of, 13–14, 68–69
Flowrate
  flow continuity and, 14–15
  flow-pressure loss and, 19
  hydraulic power and, 130–134
  maximum, 130, 189, 191, 193–194
  measurement of, 24–25
  minimum, 129, 130, 189, 191
  parasitic pressure and, 190
Fluid
  defined, 1–2
  density of, 2
  determination of properties of, 36–38
  dilatant, 19
  fundamental properties of, 1–5
  hydrostatic pressure in, 2–4
  Newtonian, 10, 12, 13, 16–17
  non-Newtonian, 10, 12, 16, 17–23, 25, 68–69
  non–time dependent, 26–27
  pressure measurement of, 24–25
  pseudoplastic, 19
  rheological properties of, 27
  rheopectic, 22
  thixotropic, 22
  types of, 1–2
Fluid deformation, 8–10
Fluid density, 2, 3
  funnel viscosity and, 25
  minimum, 129
  Reynolds number and, 13–14
Fluid flow. *See also* Flow
  analysis of, 7, 26–36
  up annulus, 7
  down drillstring, 7, 59–79
  field procedures for measuring, 38–44
  principles of, 7–15
  properties, 24–44
Fluid flow models, 16–23. *See also* Bingham fluid; Casson fluid; Herschel-Bulkley fluid; Newtonian fluid; Power Law fluid; Robertson-Stiff fluid
  field procedures and, 38–44
  fluid property determination in, 36–38
  selection of, 42–44
  shear rate in, 35–36
  shear stress in, 34–35
Fluid mechanics, 1, 24–44
Fluid velocity. *See* Velocity
Fluid viscosity. *See* Viscosity
Formation damage, minimizing, 129
Friction factor, 71–72, 121–122

Friction forces, in equilibrium, 27, 28
Funnel, Marsh, 24–25
Funnel viscosity, 25

Gas, 1
Gel strength
  drilling fluid and, 23
  excessive, 23
  fluid flow model selection and, 42–44
  fragile, 23
  progressive, 23
  in Robertson-Stiff fluid model, 22
  strong, 23
  time-dependent flow behavior and, 23
  weak, 23

Hanks, R. W., 77–78, 105–106
Hedstrom number, 77–78
Herschel-Bulkley fluid
  model of, 21–22
  pseudo, 22
  shear rate in, 36
  shear stress in, 36
Hydraulic power
  equation for, 156, 166
  flowrate and, 130–134
Hydraulics calculations, examples of, 167–194
Hydraulics program optimization, 129–134
Hydrostatic pressure, 2–4, 57

Impact force
  equation for, 156, 166
  jet, 132
Inlet coefficient, 79
Invert emulsions, 49

Jet impact force, 132
Jet velocity, 86
Johnson, M. M., 76–77

k-correction factor, 40–41
Kick, Pascal's Law and, 4–5

Laminar flow, 7
  in annulus, 91–103
  cuttings transport in, 119–120
  defined, 10–12
  field procedures for, 66–67
  fluid property measurement in, 13
  hydraulics calculations in, 172–173, 178, 179–180
  mathematical analysis of, 59–69
  in pipe, 59–69
  pressure losses in, 59–69
  Reynolds number and, 68–69
  shear stress/rate analysis and, 26–27
Laminar-to-turbulent transition. *See* Transitional flow
Liquid, 1–5. *See also* Fluid

Marsh funnel, 24–25
Mayes, T. M., 123
Mean viscosity, 68–69
Measurement, units of, 143–144

Metzner, A. B., 73–74, 76
Moineau motor, 83–84
Moment, 28–29
Moore, P. L., 122, 123
Motor, 83–84
Mud
  oil-based, 47, 49
  water-based, 47–49
  yield stress in, 17
Mud circulating system, 51–57
Mud hydraulics, calculations for, 167–194
Mud pump
  double-acting, 51–54
  single-acting, 51–54
Mud viscosity, 49–50
Mudweight, 2

Narrow annulus approximation, 96–100
Newtonian fluid, 10
  clinging constant in, 114
  field measurement of properties of, 39
  flow behavior index and, 19
  friction factor in, 72
  gel strength and, 42
  in laminar flow, 12, 62–65
  laminar flow in annulus of, 96, 97, 100
  model of, 16–17
  properties of, 36
  shear rate in, 35
  shear stress in, 34
  shear stress/shear rate relationship in, 16–17
  in turbulent flow, 13
  viscosity in, 12
Newtonian fluid model flow curve, drilling fluid curve and, 20 (fig. 2-10), 21
Nomenclature, 137–142
Non-Newtonian fluid, 10
  clinging constant in, 114–115
  flow behavior index and, 19
  friction factor in, 73
  mean viscosity of, 68–69
  models of, 16, 17–22
  shear stress/shear rate relationship in, 16, 18
  time-dependent behavior in, 22–23
  viscosity in, 12
Nozzle
  flowrate and, 134
  pressure drop at, 86–88
Nozzle size, nozzle area and, 192, 194

Oil-based muds, 47, 49
Oilfield units, 143–144
  key equations in, 145, 146, 157–166
Outlet coefficient, 79

Parallel-Plate approximation, 96
Parallel-Plate flow
  Bingham (approximate) equation for, 151–152, 162
  Bingham (exact) equation for, 150–151, 161
  Power Law equation for, 152–153, 163
Parasitic pressure, 130–134, 189–190
Pascal's Law, 4–5

Pipe flow
  Bingham (approximate) equation for, 148–149, 159
  Bingham (exact) equation for, 147–148, 158
  Power Law equation for, 149–150, 160
Pipe flow rheometer, 24
Pipe viscometer, 13
Plastic viscosity
  conversion to centipoise, 39
  defined, 17–18
Plasticity, 67, 100
Plug flow, 10
Polymers, as viscosifiers, 48–49
Positive displacement motor, 83
Power Law fluid
  in annulus, 98, 102–103
  Bingham fluid model and, 19–21
  consistency factor k in, 39–41
  field measurement of properties, 39–44
  gel strength and, 42–44
  k correction factor for, 40–41
  laminar flow in, 65, 98, 102–103
  model of, 19–20
  properties of, 37
  shear rate in, 35
  shear stress in, 34
Power Law fluid model flow curve, drilling fluid curve and, 20 (fig. 2-10), 21
Pratt, D. R., 77–78, 105–106
Pressure
  hydrostatic, 57
  parasitic, 189–190
  surge, 24, 111–117
  swab, 24, 111–117
Pressure loss, 24, 55–57
  annular, 56, 91–106
  bit and, 55–56, 87–88, 187
  calculation of, 59–79
  cuttings and, 126
  drillstring and, 55
  motor and, 83–84
  orifice and, 86–87
  parasitic, 130–134
  tool joint and, 78–79
  turbine and, 85
Pseudoplastic fluid, 19
Pump
  booster, 129
  mud, 51–54
Pump output, equation for, 146, 157

Reed, J. C., 76
Relative roughness, 71–72
Relative sinking velocity. *See* Slip velocity
Reynolds, Osborne, 68
Reynolds number, 68–69
  annular flow and, 103–104
  critical, 70
  cuttings transport and, 119–122
  equivalent, 69
  flow regime determination and, 13–14
  transitional flow and, 105

Rheological properties, of fluid, 27
Rheometer, pipe flow, 24
Rheometry
  Bingham equation for, 146, 157
  Power Law equation for, 146, 157
Rheopectic fluid, 22
Robertson-Stiff fluid
  in annulus, 98, 103
  gel strength and, 44
  laminar flow in, 65, 98, 103
  model of, 21-22
  plasticity in, 67
  properties of, 37-38
  shear rate in, 36
  shear stress in, 34-35
Rotary viscometer, 19, 22-23
Rotating sleeve viscometer, 24-36
Roughness
  absolute, 70
  relative, 71-72
Ryan, N. W., 76-77

Sample, K. J., 122
Schuh, F. J., 73, 76
Shear rate
  in Bingham plastic model, 17-18
  defined, 9
  flow time and, 22
  fluid models and, 16-22
  in laminar flow, 10-12, 62
  mathematical analysis of, 24-25, 31, 34-36, 40
  mud viscosity and, 49-50
  in Newtonian fluid model, 16-17
  in Power Law fluid model, 19-21
  shear stress and, 9-10, 16-22
Shear-rate ranges, measurement of, 25
Shear stress
  in Bingham Plastic fluid model, 17-18
  in Casson fluid model, 21-22
  defined, 8-10
  flow rate and 22
  fluid models and, 16-22
  in Herschel-Bulkley fluid model, 21
  in laminar flow, 10-12, 61-62
  mathematical analysis of, 24-43
  measurement of, 25
  in Newtonian fluid model, 16-17
  in Power Law fluid model, 19
  in Robertson-Stiff fluid model, 21
  shear rate and, 9-10, 16-22
Shear thinning, 18, 19-20
S. I. units, 143-144
  key equations in, 145-156
Slip velocity, 119-124
Sodium montmorillonite, 47-48
Specific weight, 2, 3
Sub-bentonite, 48
Surge pressure, 24, 111-117
Swab pressure, 24, 111-117

Thixotropic fluid, 22

Time-dependent behavior, in non-Newtonian fluids, 22-23
Tool joints
  equation for, 155, 165
  pressure drop at, 78-79
Torque, 28-29
Transitional flow, 13, 68-69, 76-78
  in annulus, 105
Transport ratio, 123-125
Turbine, 83, 85
Turbodrill, 83
Turbulent flow, 7
  in annulus, 103-104
  clinging constant in, 114-117
  defined, 13
  fluid property measurement in, 13
  hydraulics calculations in, 172, 176-178
  pressure losses in, 70-75
  Reynolds number and, 68-69

Velocity
  average fluid, 14, 15
  critical, 13-14
  flow continuity and, 14-15
  jet, 86
  Reynolds number and, 13-14
  slip, 119-124
Velocity profile
  of laminar flow in annulus, 91-93
  of laminar flow in pipe, 13 (fig. 2-5)
  of turbulent flow in pipe, 11 (fig. 2-4)
V-G meter, 13, 18
Viscometer
  Fann, 38-39
  rotary, 19, 22-23
  rotating sleeve, 24-36
Viscometric instruments, 12, 13, 19, 24-25
Viscometric properties
  measurement of, 24-25
Viscosifiers, polymers as, 48-49
Viscosity
  absolute, 16-17
  apparent, 18, 19, 22
  conversion to centipoise, 39
  defined, 10-12
  effective, 18
  flow time and, 22
  funnel, 25
  funnel viscosity and, 25
  mean, 68-69
  mud, 49-50
  plastic, 17-18
  Reynolds number and, 13-14
  temperature and, 44
  zero, 26-27
Viscosity gradient, in turbulent flow, 13
Viscous drag, 60-61
Volumetric efficiency, of mud pump, 51-54
Volumetric output, of mud pump, 51-54
Von Karman relationship, 74, 75 (fig. 5-5)

Walker, R. E., 123
Water-based muds, 47–49

Yield point
  in Bingham Plastic fluid model, 17
  field measurement of, 39
  in laminar flow, 61
  mud viscosity and, 49–50
Yield stress
  in Bingham Plastic fluid model, 17
  in fluid models, 17, 19–21
  in Herschel-Bulkley fluid model, 22